No. 497

TAPE RECORDING FOR FUN & PROFIT

By Walter G. Salm

TAB BOOKS
BLUE RIDGE SUMMIT, PA. 17214

FIRST EDITION

FIRST PRINTING — NOVEMBER 1969

Printed in the United States
of America

Library of Congress Card Number: 76-77379

Preface

Not very long ago anyone who wanted to make a recording at home had to spin a shellac disc at 78 RPM on a turntable of questionable reliability, using a cutting stylus that was forever wearing out. Occasionally, someone got a shellac chip in his eye—provided somebody else remembered to blow it away from the platter during the recording session! And all this for three minutes of a static-filled home-brew record with incredibly bad fidelity.

After two decades of practical tape recording equipment availability on the American market, the "revolution" is really still in its infancy. Today, there's hardly a household in America that hasn't owned at least one tape recorder, although many of these machines have been the dime-store variety and have long since ended up in the junk heap. But to some of us, the tape recorder has become a way of life, a valuable business tool, a highly versatile source of enjoyment and relaxation, a study aid, and generally a third hand in many activities, both indoors and out.

The advent of economical battery-operated portable recorders has made tape even more versatile and useful. Trouble is, many companies are peddling their own systems and their own standards, and the layman who doesn't know his way around needs some kind of a scorecard to tell him what's what. We do not pretend to have all the answers to all of the questions about tape recording, but we try to satisfy some of them, regarding differences among tape formats, the tape marketplace, as well as how to get the most from your equipment. We certainly hope you'll come away with some useful ideas for getting more versatility and enjoyment out of your tape recorder.

Walter G. Salm

CONTENTS

1 **How Tape Recorders Work** 7

Natural Magnets — Flux Density — Making the Recording — Different Kinds of Plastics — Oxide Formulations

2 **Scorecard For Tape Formats** 16

Endless Loops — What's a Cassette? — No Accidental Erasures

3 **Choosing A Tape Recorder** 26

Batteries vs High Fidelity — Why a Deck? — Index Counters — Automatic Shutoffs — Automatic Reversing — The Seconds of Silence — Solenoids or Belts and Pulleys — Transistor vs Tubes — Level Indicators

4 **Battery-Operated Recorders** 43

Primary and Secondary Batteries — The Most Common Battery — Alkaline Primaries — The Mercury Cell — Shelf Life — Secondary Batteries — Recharging Primaries

5 **Microphone Basics** 56

Mixed Bag of Types — Enter the Condenser — Deciphering the Spec Sheet — Pickup Patterns — The Most Useful Pattern — Getting the Right Impedance — Output Level — What About Those Wires?

6 **Getting The Most From Your Mic** 75

Know Your Gear — Polar Pattern — Windscreens — Proper Phasing — Post-Session Mixing — Microphone Accessories — Add a Boom — Mixing Mikes — Recording in Stereo — Center Fill — Room Acoustics — Some Dos and Don'ts

7 **Wireless Microphones** 95

Building a Wireless Mike — Frequency Response — The Versatile FM Mike

8 **Amplifiers And Preamps** 103

Inexpensive Preamp – Integrated Circuit Preamp

9 **Microphone Mixers** 115

Amplifier Mixers — Low-Cost Units

10 Remote Control & VOXs 123
Voice-Operated Switch — VOX With AC Machines

11 Switching Adds Versatility 129
Making the Hookup — Building the Stereo Switchbox — Buffer Amplifier — Clean Sound

12 Plugs, Jacks & Cables 137
Cable Types — Don't Always Use Coax — The Phono Plug — The Phono Jack — Phone Plugs — Japanese Plugs — Screw-On Plugs — European Plugs

13 Integration With A Stereo System 150
Adding Connectors — Adapting For Playback — Component Systems

14 Recording From A Telephone 158
Pickup Tubes — Line Preamp — Recording Speed

15 Recording Film Sound Tracks 167
Sound Stripes — Dubbing the Sound Track — Instant Sound Tracks — Tape Soundtrack — Automatic Sync — Sound For Slides

16 Maintaining Tapes & Recorders 175
Splicing and Leaders — The Splicing Block — Routine Maintenance — Cleaning the Heads — Degaussing the Heads — Transport Troubles

17 Special Effects In Recording 185
Basic Reverberation Unit — Tape Reverb — Center Mike Fill — Panpot Control — Tape Editing — Finding the Spot

18 Foreign Recorders & The Future 195
The Newell Transport — High Speed Tapes

19 How Tapes Are Duplicated 199
The Master Tape — Cartridges and Cassettes — The Hiss-Killer — How It Works — Making the Master

20 Video Tape Recording 207
Bandwidth Needed For Video — In-Line Tracks — Helical Scan Machines — Omega Wrap — Missing Interlace — Linear Scan — Video Tape In the Home — Service Problems

21 The Snoopers Around Us 216
Tape Recorder Is the Key — VOX Circuits — Long-Distance Bug — Russian Bugs — The "Anti-Bug"

Chapter 1

How Tape Recorders Work

Since you are reading this book, chances are you've used a tape recorder at one time or another. You know that by turning the machine on, pressing a button, and talking into a microphone that some sounds will issue from the loudspeaker when you replay the tape. What you hear may or may not sound very much like your voice, depending on the machine, microphone, and your own appraisal of what you should sound like. But somehow some voice patterns have gotten onto that skinny little reddish-brown ribbon of plastic and somehow or other they can be reproduced and erased if needed. Let's take some of the mystery out of this process right now.

NATURAL MAGNETS

In the beginning, there were just loadstones which were really just magnetized hunks of iron ore. Now there are lots of materials that have magnetic properties besides iron, but iron is the most obvious and readily available magnetic mineral. Magnetism still isn't fully understood, but it involves a kind of alignment of small particles or groupings of molecules within the iron (see Fig. 1-1). These particles are called "domains," and each domain can contain several dozen molecules of iron or iron oxide. Each domain is like a tiny magnet with a north and south pole, and when a bar of iron is magnetized the domains line up north to south (opposite poles attract) all across the piece of iron so that the large bar itself will also have a north and south pole (see Fig. 1-2). In a permanent magnet made of steel or alloy, these domains remain aligned in this position. In a magnetized piece of soft

Fig. 1-1. The little bars in this drawing represent the random arrangement of domains in unmagnetized iron oxide or other magnetic material. Each domain has a "north" and "south" pole.

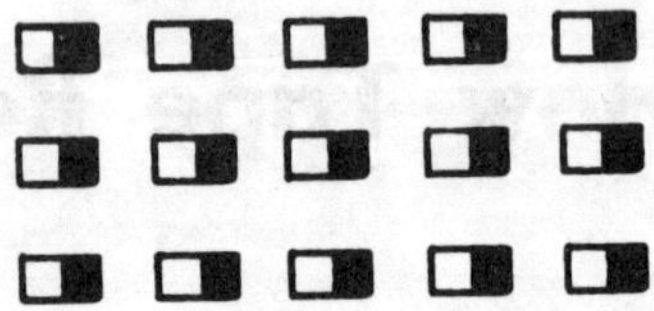

Fig. 1-2. Here, the magnetic domains line up in magnetized iron oxide or other magnetic material—typically a permanent magnet.

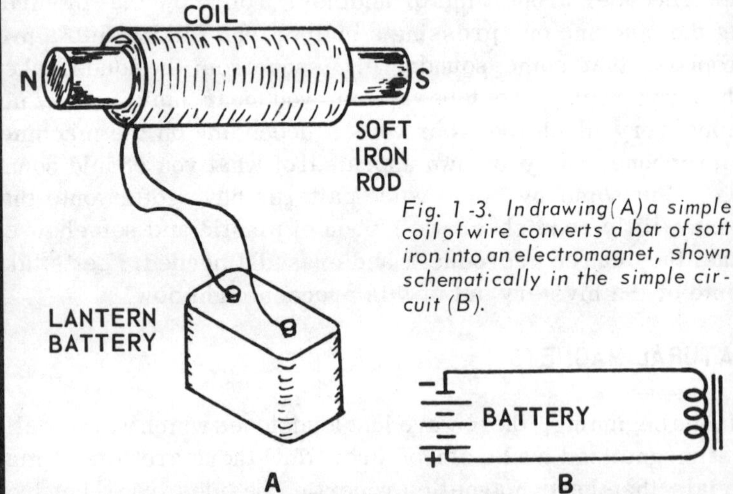

Fig. 1-3. In drawing (A) a simple coil of wire converts a bar of soft iron into an electromagnet, shown schematically in the simple circuit (B).

iron, they stay this way only while held there by the presence of an external magnetic field.

If we wrap a coil of wire around a rod of soft iron and connect it to a battery, we have an <u>electromagnet</u>. As long as electric current is flowing through the wire, the bar of iron is magnetized (see Fig. 1-3). When the current is interrupted by opening a switch, the bar is no longer magnetized.

The basis for this action does not depend so much on the magnetic properties of iron itself but on the fact that magnetism is the very basic stuff of electricity. An electric current is the motion of electrons through a conductor (copper, iron, most metals, actually). Every one of these tiny electrons has a magnetic field around it, and when the electron moves the magnetic field goes with it. What the iron rod in an electromagnet does is to concentrate this magnetic field in one specific area.

FLUX DENSITY

If we bend the iron rod (of the electromagnet) into a horseshoe shape, we still have an electromagnet, but now one with a magnetic flux field that is highly concentrated in one area—between the two poles (see Fig. 1-4). If we move the two poles closer and closer together, the flux density becomes much higher; that is, the flux density—or magnetic force—available at a given point is much greater.

If we vary the electric current passing through a coil of wire, the magnetic field around the coil will vary in much the same way. If the coil of wire is wrapped around a piece of soft iron, the magnetic field in the magnet will also vary in much the same way (see Fig. 1-5). A stronger electric current will result in a stronger magnetic field; a weaker electric current, a weaker field and so on. In the same way, the rate of frequency or current variation (per second usually) is reflected by a varying magnetic field.

Fig. 1-4. This small, permanent horseshoe Alnico magnet is so powerful, it can lift over five times its own weight.

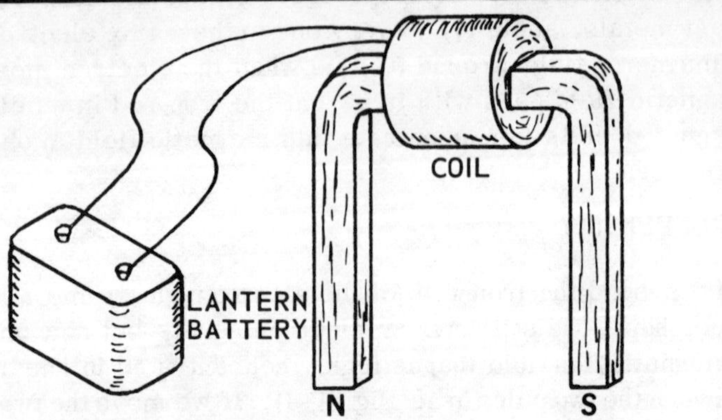

Fig. 1-5. The coil of wire converts this horseshoe-shaped piece of soft iron into an electromagnet when a current is passed through it.

Fig. 1-6. The heads on a tape recorder are usually concealed by a protective plastic cover. Here, the cover's removed for cleaning, exposing the heads to view.

The horseshoe-shaped electromagnet is the basic electronic device in a tape recorder. And in a tape recorder, it's called the "head." The tape recorder head is naturally much more refined than a horseshoe magnet. The pole pieces are brought extremely close together so that a microscopically thin line separates them (see Fig. 1-6). This thin line is called the "gap," and the tape recorder's frequency response largely depends on the size of this gap—the smaller the gap, the higher the frequency response—up to a point.

MAKING THE RECORDING

Fine, you say, but how does this record a sound on tape? Well, let's see what we can do with the magnetic field that's around the gap. If we take a piece of steel ribbon or tape and move it across the head gap while varying the voltage, we'll vary the alignment and arrangement of the magnetic domains on the tape. This rearrangement represents the "recording." Since steel has high magnetic permeability, it will retain the magnetic image in its domains after it's out of the magnetic field—provided the field is strong enough.

This field can be made to vary in "step" with sounds picked up by a microphone and boosted by an amplifier circuit, which in turn feeds fluctuating voltage to the head (see Fig. 1-7A). The amplifier is necessary for the simple reason that the microphone alone cannot put out a strong enough signal voltage to drive the recording head. Also, the signal probably

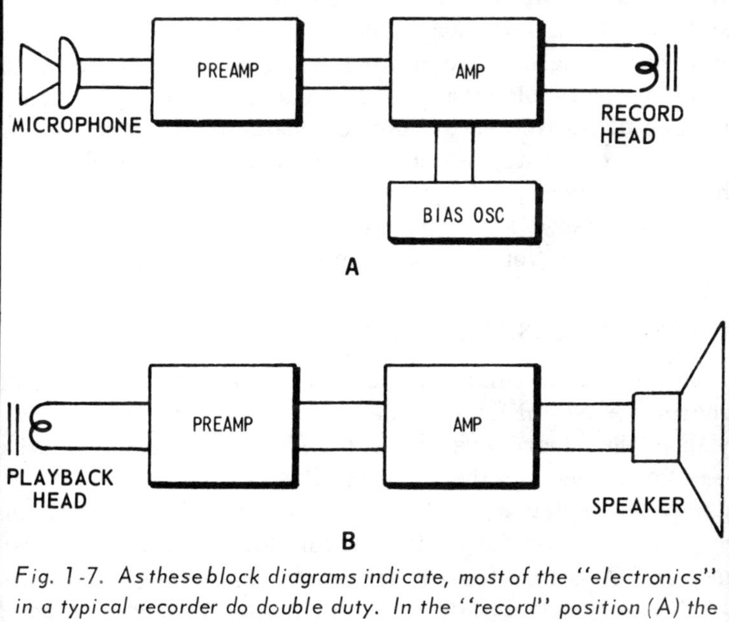

Fig. 1-7. As these block diagrams indicate, most of the "electronics" in a typical recorder do double duty. In the "record" position (A) the microphone is connected to the preamp; in "playback" the tape head is switched to the preamp input (B).

would lack certain other important characteristics, such as proper equalization.

Once these variations in the pattern of the steel tape's domains have been made, playback is simply the reverse of the recording procedure. This time the "clumps" or clusters of domains in the tape generate a feeble magnetic field in the head. This in turn produces a tiny electric current in the head coil which goes through an amplifier circuit and then to a loudspeaker (see Fig. 1-7B).

Modern tape recorders generally use a plastic tape coated on one side with iron oxide instead of steel tape. The iron oxide is the recording medium and the plastic is the base material—just as important as the oxide if not more so. A tape is usually classified by the characteristics of the plastic base itself. Thus, we have acetate tapes, mylar tapes, polyester tapes (another name for mylar) and various other names. Special processes are used in preparing different kinds and grades of tapes, resulting in a wide spread in retail prices.

In manufacturing tape, iron oxide is ground into very fine particles by a ball mill. One of the biggest problems in manufacturing is maintaining a close tolerance on particle size and then coating the tape uniformly with these particles without clumping or blank spots. Any such clumping or bunching together of particles, or any low spots in the oxide coating, can cause dropouts, which are instantaneous signal losses. This could mean a missed note in a musical piece, or much worse, a missed key word in a speech or number in a long string of calculations in a computer.

DIFFERENT KINDS OF PLASTICS

Why not standardize on one type of plastic, you may ask. There are two different plastics used in tapes for the same reason that there are several different oxide formulations —not everyone has the same requirements.

Acetate, which is the older of the two, is cheaper than polyester (mylar) but its long-term storage properties aren't as good. Acetate's tensile strength is relatively low, but when the tape breaks, it makes a clean snap with no stretching. This is both good and bad. It's good because the tape won't stretch—which could cause some frequencies to change by lengthening the time a given note takes to pass over the playback head. It's bad because acetate tends to break more frequently than polyester. The oxide also tends to chip off the plastic base, especially at the edges.

Polyester has high tensile strength. It'll stretch way out of shape long before it will break and it will absorb mechanical shocks and stresses more easily than acetate tape. There's no known lifetime limit on polyester's storage properties. On the other hand, acetate starts to lose some of its magnetic properties after 10 years or so; therefore, the material recorded on it deteriorates.

Because of its high tensile strength, polyester tape can be made very thin—giving rise to the "extra-play" and "long-play" tapes that are available from just about every tape manufacturer. There are generally three different tape thicknesses: 1.5 mils, 1.0 mil, and 0.5 mil. The last one is used for the double- and triple-length tapes, since more tape can be wound on one reel.

Because triple-play tape is so thin, it's often hard to handle; ordinary static electricity (of which there's always plenty around) makes the tape climb your fingers and obstinately refuse to go between the tape recorder's guides and onto the

Fig. 1-8. White leader tape may be spliced onto a reel of tape for protection, indentification, and timing.

takeup reel. There's only one thing to do with tape such as this: splice "leader" and "trailer" tape onto it (see Fig 1-8).

Leader tape is made of polyester with no oxide coating, and it's treated to give it some additional strength. You can buy it in white, yellow, green, or a variety of other colors. Often, it's marked off in 3 3/4-inch lengths, for timing purposes. "Scotch" brand leader and timing tape is marked off with a little square tartan plaid every 3 3/4 inches for this purpose.

OXIDE FORMULATIONS

There are several oxide formulations presently used on tapes. Cheapest is the standard, utility grade tape with no fancy names on it, no matter who the manufacturer. Then there's the wide-range low-noise tape that has lots to commend it. This tape costs a bit more than utility grade, but it makes a big difference in the sound. It works quite well at slow speeds and with average-to-good equipment will give excellent frequency response, dynamic range, and signal-to-noise ratio at 3 3/4 ips. This way, the tape cost is actually cut in half for the same playing time, so the premium price is actually quite cheaper overall.

Then there's "low-print" tape. "Print-through" is a perennial problem, especially with thin-base tapes and tapes that are going to be stored for a long period of time. Here's what happens: When there's a loud passage of music on the tape, the magnetic field of the tape domains is so strong that it affects the domains in adjacent tape layers—both above it and below it — on the reel. When this happens, the loud passage will have an "echo" — both before (pre-echo) and after (post-echo) the loud passage itself. This defect is called print-through.

Low-print tapes are thick (1.5 mils); therefore, there's more space between oxide layers. Also, to further lessen the chance of print-through the formulation may not have such a wide dynamic range. Ideally, such a tape should be very hard to record — so the domains are less susceptible to print—but this isn't always practical.

The print-through problem is so severe that the British Broadcasting Corp. copies all of its priceless tapes on discs and stores both the discs and tapes in its archives. There is a way of eliminating print-through electronically, but the equipment involved is frightfully expensive for the hobbyist. Besides, there's such a waiting list for this gear that even major recording companies and broadcasters have to wait months for delivery. This equipment—known as the Dolby noise reduction system—is described in Chapter 19.

Chapter 2

Scorecard For Tape Formats

There are many different "formats." Some are no longer in popular use, while others compete with each other for the top spot in the marketplace. You're probably familiar with the standard reel-to-reel format (see Fig. 2-1). This is the "original," and it continues to be the most widely used method of tape packaging. Tape comes in other packages, too—mostly cartridges of some kind that eliminate tape threading.

The basic idea behind eliminating threading is to encourage people to use tape—especially the "all-thumbs set" who normally would be afraid to tangle with that red oxide ribbon. Of the cartridge type systems available today there are

Fig. 2-1. The open-reel tape is the most common format and the most versatile for general-purpose and high-fidelity use.

Fig. 2-2. The enclosed RCA cartridge, although cumbersome and restricted, was the forerunner of today's cassette systems.

four formats in competition: the 4- and 8-track endless loop cartridges; the compact cassette; the endless loop 2-track Playtape. All have one common denominator—they try to make tape as easy to use as records on a phonograph.

An early system that made a valiant effort was introduced a number of years ago (see Fig. 2-2). It used a large cartridge by today's standards, basically a reel-to-reel concept all enclosed within the cartridge. It worked fine, but never caught on. It was bulkier than a standard 7-inch reel of tape and appreciably more expensive. Early proponents of the system dropped by the wayside as sales flagged. RCA tried to go it alone for a few more years but finally bowed out in the face of competition from the compact cassette.

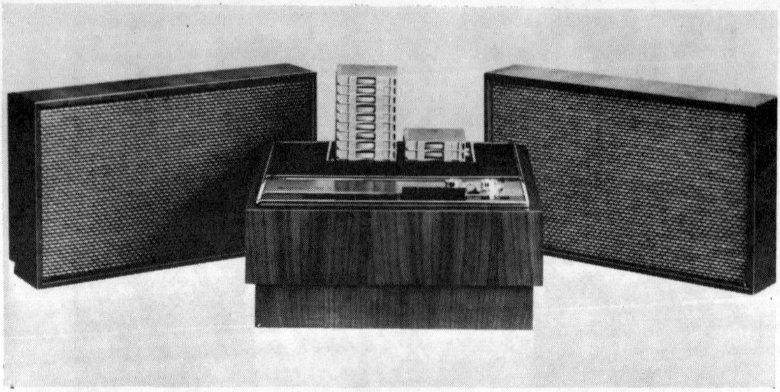

Fig. 2-3. The 3M cartridge system depended on a complex changing mechanism and a special type of cartridge. Like the RCA cartridge recorder, this idea also withered on the vine.

Another cartridge system that has withered on the vine is the cartridge changer invented by Wollensak. This changer used a self-contained prerecorded tape that could be stacked along with many others in a changer mechanism (see Fig. 2-3). The tape machine would thread the tape automatically and play it through in one direction only. Then the tape would be rewound into the plastic shell automatically. When the rewinding was finished, the cartridge would be ejected and another tape would fall into place and start a new cycle. This system was beautifully simple to operate but incredibly complicated as far as the machinery went. Again, this system has been supplanted by more viable and economically feasible ones and has since slipped into oblivion.

ENDLESS LOOPS

The 4-track continuous loop cartridge is the oldest such system still in use today. It uses a standard 1/4-inch wide audio tape that is extremely well lubricated. The tape is wound around a single hub and as it plays the tape unwinds from the center of the spool (see Fig. 2-4). The tape passes through a system of guides, over a pair of pressure pads (which hold it against the playback head) between the pinch roller and capstan, and back into the maw of the cartridge where it winds onto the outside of the spool.

As the tape plays, successive layers slide over each other,

each layer constantly changing its diameter as it works its way towards the center of the hub. The amount of friction involved is enormous; that's why the tape must be very well lubricated. This constant sliding action is a frequent source of wow in the tape playback system—a shortcoming that is certainly understandable.

The tape plays two stereo programs, each roughly equivalent to one side of an LP record. The head assembly simply moves down to play the second stereo pair. In many players the head is moved by a manual control lever at the end of the first program, but in some systems the head moves automatically—triggered by a strip of metal on the tape itself. Such systems are more expensive, obviously.

The direct descendant of the 4-track cartridge—the 8-track continuous loop cartridge—operates essentially the same way except that it has twice as many tracks within the given width of 1/4-inch tape (see Fig. 2-5). That means the track widths are narrower—less than half the width of those on 4-track tapes. Hi-fi buffs can wag their fingers and say that this means poorer frequency response and poor signal-to-noise ratio. This may be true, but these factors certainly don't interfere with the tape's intended purpose for highway listening. Then too, in a car there's plenty of road noise to mask any background noise problems on the tape.

Fig. 2-4. The 4-track or "Fidelipac" cartridge has a hole on the bottom to admit playback machine's "pop-up" pinch roller.

Fig. 2-5. The 8-track (Lear-Jet) cartridge has its own rubber pinch roller built in (at the right of the cartridge in this photo).

The 8-track cartridge also suffers severe wow problems, especially after the lubricant on the tape has begun to age somewhat. The 8-track cartridge's vast popularity is due to original equipment offerings by manufacturers. Buy a car and have an 8-track stereo cartridge player installed for only X number of dollars extra—it all sounds very enticing and very plush, and very much "with it," so naturally the prospective new car buyer goes for it. After all, Mr. Jones next door already has one.

Another problem with the 4- and 8-track cartridges is that they're hard to record at home. There are "home" machines that will make such recordings, but the resulting quality often leaves something to be desired, and the cost is generally too high, even with available inexpensive blank tape cartridges. Typically, a 150-foot length of tape in an 8-track format can cost $3.00—this for about 32 minutes of playing time.

In operation, the 8-track playback head moves down to the next pair of tracks at the end of the first program. It's motion is triggered by a metal strip on the tape. It's endless music, until you reach down and pull the cartridge out of the machine which shuts the system down.

A couple of other basic differences between the 4- and 8-

track cartridges; The 8-track cartridge system has its own pinch roller built in; the 4-track cartridge doesn't; instead, it has a large hole cut out of it to accept a pinch roller from the playback machine. This roller is usually lying on its side when the cartridge is inserted. When the playback lever is pressed, the roller jumps up through the hole and locks itself into place. The 4-track cartridge also has a ratchet interlock which prevents the tape from unraveling or unwinding when the unit is not actually in an operating position in the machine.

The Playtape is another continuous loop playback-only format, this one aimed strictly at the teenage set. It's a tiny continuous-loop cartridge with only two tracks of monophonic music on it. Tracks are selected by a control switch on the player. The bottom-of-the-line playback machine—about 2/3 the size of a cigar box—sells for \$19.95. Playtape cartridges cost about \$1.50 each and generally contain about

Fig. 2-6. Revealed here are the insides of an 8-track cartridge (l), a 4-track cartridge (c), and a Philips-type cassette. Notice the comparative size of the cassette.

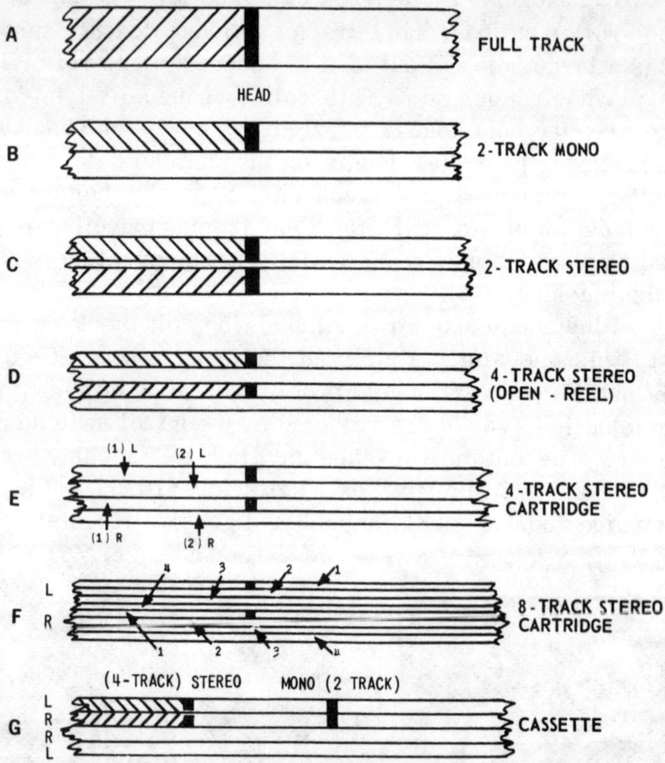

Fig. 2-7. Represented here are the various tape formats in current use. Full-track is classic method (A) using the full width of tape for single channel. Two-track mono (B) was developed mainly for home tape recorders when tape formulations and heads had been improved enough to make it practical. Two-track stereo (C) is still used for masters in recording studios and broadcast stations. Four-track on open-reel tapes (D) uses alternate tracks for left and right channels, permitting the tape to be turned over for a second side of music. The four-track cartridge (E) always plays in the same direction, using two pairs of tracks for the two programs. The eight-track cartridge (F) has four stereo pairs for four programs of about seven to ten minutes each. The cassette (G) has left and right tracks next to each other, making the stereo cassette fully compatible with a mono two-track recorder.

four popular tunes. More complex playback machines are available—even automobile installations—but Playtape remains mostly a teenage phenomenon.

WHAT'S A CASSETTE?

Probably the most far-reaching development in the field was the invention of the compact cassette by N.V. Philips Gloeilampenfabrieken in 1963. It made its appearance at the New York High-Fidelity Show that year and was an instant success. The original concept has been carried through in subsequent equipment design changes, while the cassette itself has remained the same. It's now licensed to about 60 other companies and is truly a worldwide standard.

The cassette is tiny. It uses a basic reel-to-reel approach for the tape, eliminating the friction associated with the continuous-loop. Instead, the tape that is wound on the take-up hub occupies space just vacated by the tape unwinding from the feed hub. The tape itself is about 1/7 inch wide and can contain up to four tracks of information when used for stereo. The standard tape speed is 1 7/8 ips (inches per second). Generally, a low-noise formula tape is used.

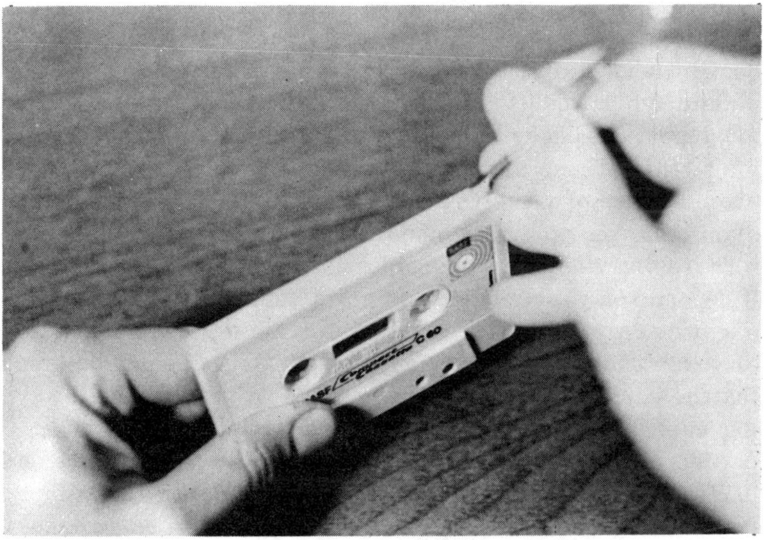

Fig. 2-8. As shown here, a tab on the rear of a cassette can be pried loose to prevent accidental erasure—making the cassette pretty much "goof-proof."

In its original design concept, the cassette was used with a small cigar-box size battery-operated portable tape recorder capable of 2-track monophonic recording. The little "Carrycorder" sold for $150 when it was first unveiled and now has a price tag of $50 to $60. Many Japanese companies and some other European firms also offer their versions of this machine for the Philips cassette at similar prices. Cassettes can be purchased as prerecorded stereo albums holding up to 120 minutes, compared to the theoretical maximum of 80 minutes on an 8-track cartridge.

The stereo cassette is fully compatible with its monophonic counterpart. Instead of having alternating heads or tracks for the stereo pair (as in other stereo tape systems), the cassette has the stereo pair side by side. Thus, a 2-track playback head covers both the left and right channels, at the same time playing back a full monophonic program. Similarly, a monophonic cassette can be played on a stereo machine with no interchangeability problems.

NO ACCIDENTAL ERASURES

Another feature of the cassette is its anti-erase interlock, a feature found on no other tape system in use today. If you've got something on a cassette that you don't want to erase accidentally, you just take a penknife and pry a little plastic tab in the back of the cassette. From that point on, the tape is erase-proof. At some time in the future, you may want to reuse the tape. When that time comes, simply place a piece of adhesive tape over the hole uncovered by the plastic tab and you're in business again.

There are many imponderables in selecting a tape format. It depends on your intended primary use. If it's going to be playback only, then you would do well to consider a cassette or a cartridge system. Many industry spokesmen believe that the cartridge is on its way out and will be supplanted by the cassette. Certainly there are at least as many prerecorded albums available on cassettes now as there are on cartridges, and the cassette is much smaller and more convenient. Four cassettes can fit into the same space taken by a single 8-track cartridge, each cassette holding more information also (see Fig. 2-9). Also, the cassette can be recorded very easily. You can "roll your own" stereo cas-

Fig. 2-9. If space is a problem, four cassettes fit easily into a case large enough to hold a single eight-track stereo cartridge. Total maximum playing time for the big cartridge: about 80 minutes; total for the four cassettes: 480 minutes.

settes at home and play them back in the car. And, you can use the unit in your car as a dictating machine and recorder for jotting down those important notes as you cruise along the highway.

The ultimate choice depends on your needs. If you expect to make true high-fidelity recordings, then obviously open-reel is your choice. If you want compactness and versatility, the cassette may be the one that you want. If you want to go along with the crowd and not have to worry about doing any installation in your car, then go for an 8-track cartridge since you can have that installed in the car when you buy it brand new—but that particular system has built-in obsolescence. Just bear that in mind.

Chapter 3

Choosing a Tape Recorder

Once you've decided upon a tape format, there are other decisions to make. First of all, do you want mono or stereo? Do you want an AC or a battery portable machine? Do you want a combination of both? If it's an open-reel machine, what size reel should it accommodate? How many speeds should it have? What frequency response do you need? How good must the wow and flutter characteristics be? Do you need an index counter? What about automatic shutoffs? Is a pressure pad system acceptable or do you want the professional tape tension system? What about automatic reverse? The list goes on and on.

There are many factors that you have to think of when you pick a tape recorder. Let's start at the top of the list. First off, there's no question that if you need superior frequency response, you are going to have to get an open-reel machine. Whether or not this machine is battery- or AC-operated has little bearing on the frequency response with today's technology. Most of the current Japanese battery portables do not provide as good a potential frequency response as most AC machines do.

BATTERIES vs HIGH FIDELITY

You'll find that there are a few—very few—battery operated portable recorders that give superb frequency response (see Fig. 3-1). Most of these machines take only a 5-inch reel for the sake of compactness. Don't expect to find all of the features you want in the battery machine plus 7-inch reel capacity. There's only one (expensive) exception that we know of (see Fig. 3-2).

Fig. 3-1. Typical of today's breed of quality battery/AC recorders imported from Japan, this Telmar unit takes a 5-inch reel, has remote control and VOX control option.

If you want a machine to be an all-around expert, with plenty of versatility, what you need is a 7-inch reel, AC machine that folds up into a compact package (see Fig. 3-3). A machine like this ideally should have: a usable frequency response from 50 to 15,000 Hertz at the 7 1/2-inch per second speed; an automatic shutoff; two built-in speakers (probably of poor quality); enough amplifier output to drive auxiliary speakers if needed; a clutch assembly which disengages the capstan pinch roller when the machine shuts off; an instantly resettable index counter; and some kind of pause control. The machine should be capable of being played in a horizontal or vertical position. Additional features which would be desirable are remote control, VOX (voice-actuated control), a selection of at least three speeds (7 1/2, 3 3/4, and 7/8 ips), separate tone controls for each channel, separate recording and playback level controls for each channel, a record safety interlock that really works, and VU meters for recording level indicators.

This description lets out quite a few tape recorders that might otherwise be on your list of potentials. If you want to narrow it down further, you must start setting price criteria and must be willing to hedge a little on some of your requirements. Remember, this tape recorder will be used as a portable machine, it will be used as a music source at parties, it will be used for recording off the air and from records, it will be used for recording speech and dictation, it will be a source of party fun and possibly for making your own movie soundtracks. Its other potential uses are so numerous and so far-reaching that it's impossible to go into them here. The amount of use that you get out of your tape recorder is limited only by your own ingenuity and imagination.

Fig. 3-2. This high-quality Concertone stereo recorder will work on AC or battery supply, making possible excellent recordings in the field. It has 7-inch reel capacity.

WHY A DECK?

If you want to go into a somewhat more specialized machine, you may want to consider a deck. A deck is simply a tape recorder stripped of its audio amplifiers and loudspeakers. It must be cabled into a stereo music system for playback (see Fig. 3-4). It's not intended to be portable; it's meant to be built-in and stay in one place.

A deck has many advantages over a portable tape recorder. For one thing, the deck version of the same tape recorder will be cheaper, since you're not paying for speakers and amplifiers. Another thing: Once installed it will not be subjected to constant plugging in and unplugging with the resulting wear on cables, intermittent connections, and other

Fig. 3-3. Compact and rugged, this discontinued tube-type recorder by 3M (Model T1580) was an early full stereo entry and is still in wide use today.

AMPEX 2100

Fig. 3-4. A professional-quality deck, the Heathkit AD-16 (top left) is a build-your-own version of the Magnecord 1020. The Ampex 2100 deck (bottom left) is a versatile machine with automatic reversing to make listening a breeze. The Bell & Howell Autoload (above) threads itself and has flip-over heads for playing the other side of the tape.

problems that you will have when a tape recorder is only sometimes part of your stereo system. As a permanent fixture in your stereo system, you will come to know your deck so well and so completely, it'll be an easy matter to do the most difficult taping jobs. But the deck has no legs. If you want to do some recording on location, you'll either have to get a portable recorder, or build some kind of a case for your deck so you can slip it out of its moorings and take it on the job when the occasion demands it.

The ideal home system uses two decks—one for playback and the other for recording. With this kind of setup, you can purchase equipment that is optimized for each function. You'll also be in the wonderful position of being able to "dub" from one tape to another and to edit tapes electronically.

Fig. 3-5. A basic cassette recorder, the 3M Model 4200 has a host of accessories to make it more versatile. This portable unit is 2-track mono but will play stereo cassettes.

If you want to go the cassette route, an optimum system would include three basic machines: a mono portable cassette recorder (see Fig. 3-5), a mono portable recorder/playback unit for home and beach (see Fig. 3-6), and a 4-track stereo playback and mono record machine for your car (see Fig. 3-7). By setting up the system with three machines, you can get versatility by using the portable for on-the-go dictation or on-the-go music. The little mini portable is great for beach parties, picnics, and so on where some music would certainly be appreciated. The automobile recorder provides the option of dictating as you drive or just throwing in some music cassettes.

The home system is the root and hub of the entire combination, since this is where you record your own cassettes and where you get the maximum amount of playback potential from these tapes. The home unit can be a deck wired into your major stereo system, in which case the cost of the recorder is minimal. A very inexpensive play-only mono portable can also make the scene at beach parties and such.

INDEX COUNTERS

Index counters are a must for most recording sessions. The home cassette recorder will have an index counter, as will most open-reel machines. The index counter does not

Fig. 3-6. Anyone for the beach? This mono portable by Ampex plays mono or stereo cassettes. It has a larger speaker for fuller sound quality.

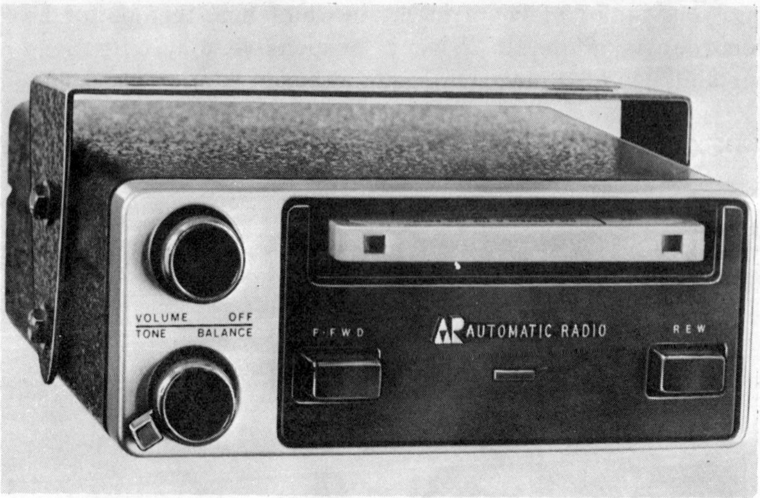

Fig. 3-7. This cassette tape player for the car by Automatic Radio fits under the dashboard. It has "push-in" loading.

show the number of inches or feet of tape length of playing time. What it does show is the number of revolutions made by the feed reel (see Fig. 3-8). The number of revolutions per inch or per foot of tape varies with the diameter of the remaining amount of tape on the reel, so this number represents nothing except a convenient way of finding a specific spot on the tape. Chances are the index counters of two different machines will not even read the same for the same number of revolutions, so when you write down those numbers just remember which machine you got them from. It'll come in handy.

AUTOMATIC SHUTOFFS

There are various kinds of automatic shutoffs. The most common type is simply a feeler under the tape that senses the end of the tape; when the end of the tape passes, the feeler moves and a switch opens, shutting off electricity to the machine (see Fig. 3-9). This system is good for one reason: it shuts off after you've set the machine on fast forward or fast rewind, not just after the end of a program.

This system's main shortcoming occurs in those machines that don't have an automatic clutch, in which case the ma-

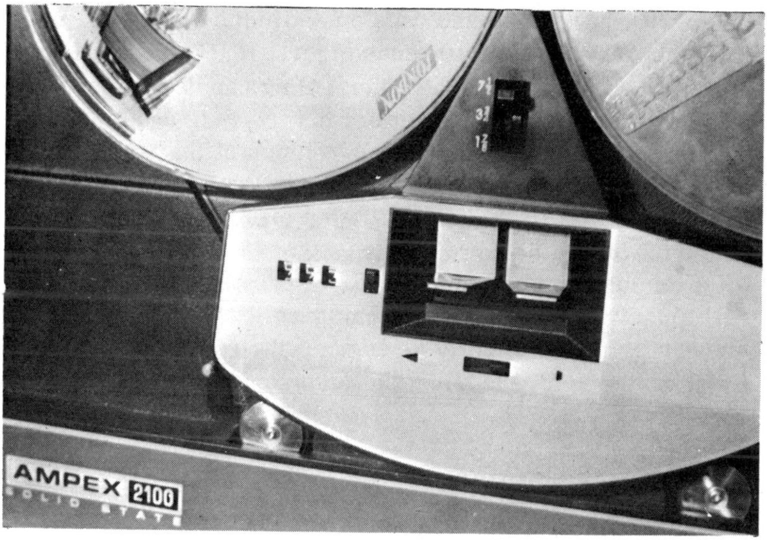

Fig. 3-8. A digital index counter helps determine the approximate location of a desired selection on the tape; it should have a push-type instant-reset button as shown here. Contrary to popular belief, the counter doesn't measure time, tape length, or anything else except the number of revolutions made by the feed reel.

Fig. 3-9. One type of automatic shutoff uses a tape guidepost that is spring-mounted and attached to a snap-action switch.

chine will shut off with the capstan and pinch roller still making intimate contact with each other. If left for any length of time in this position, the pinch roller can develop a "flat" which will eventually cause flutter in recording and playback.

Some systems sense the lack of tension of the tape passing across movable guideposts on either side of the head assembly. This type of system will shut off automatically at the end of the tape in record or playback modes, but will not shut off after fast forward or rewind. Here, there's no problem with possible "flats" on the pinch roller, since the clutch mechanism automatically disengages the roller and capstan. But it would be nice to have that auto shutoff for fast winding speeds.

AUTOMATIC REVERSING

And what about these newer machines with automatic reversing? Auto reverse is a very desirable feature, and having it available on the tape playback machine has made many of us very lazy. It's the height of luxury to be able to put on an album-length reel and not have to turn it over halfway through (rethreading tape) and starting it all over again!

With an auto reverse machine, it's a simple matter to put on the tape and drop it into an easy-threading pickup reel. Then just sit back, relax, and forget about any more mechanical functions. When the tape reaches the end of the first side, it automatically reverses and plays in the opposite direction until it's completely finished. Then the machine shuts off. It's the same effect as though the machine were to play first one and then the other side of an LP record, all automatically, with no help from the listener.

There are many machines on the market today that have automatic reverse, and there are four distinctly different systems for triggering this automatic feature. Each one has its merits; each one has its flaws. The system that you ultimately choose will depend not only on the reversing system but on the merits of the machine itself.

Probably the best known triggering system for reversing tapes is a metal foil strip; it's also the most widely used. To make this system reverse the tape at the right spot, the hobbyist must paste a short strip of metal foil to the recording tape (see Fig. 3-10). The foil is purchased in a separate spool at a modest price. After a small amount of practice, you can become pretty adept at putting the foil onto the tape with the first try.

In operation, when the metal foil passes over two finger-like contacts, it triggers a solenoid or relay in the machine, causing the motor or the pulley to reverse direction. It also changes the connections to the heads so that another set of heads is now in the system. The tape plays in the opposite direction as easily as it did in the first. One tape system (Bell and Howell) uses the same heads during both directions of tape travel. Here, the head is mechanically flopped over to the new position.

A less widely used but very convenient end-of-tape sensing system is employed in all automatic Ampex machines. This system uses a 20-Hertz audio tone recorded on the tape. When the playback circuits receive this tone, the machine reverses automatically. This tone is placed on the tape very easily by pressing a lever on the recorder. In playback, the machine will dutifully reverse when this tone is reproduced. The tone can be placed at both ends of the tape if desired, and the same tape can be made to repeat endlessly. On pre-

Fig. 3-10. Metal foil must be added to the tape to actuate the automatic reversing mechanism in a vast majority of auto-reversing recorders.

recorded tapes, Ampex includes this 20-Hertz tone (free of charge).

THE SECONDS OF SILENCE

Another audio reversing system, again used by just one company, is the Sony silence-sensing technique. This method relies on a period of 10 seconds of silence in all four tracks. When this happens, the machine reverses automatically. The only problem here is that the machine can't distinguish between 10 seconds of silence programmed into a prerecorded tape and the "real McCoy."

The fourth reversing method uses a trapped leader (or trailer to be more precise) on the tape. This is a piece of leader (trailer) tape at the end of the first side that's attached to itself so it can't pull off the feed reel. When the tape recorder senses the increased tension on the tape at the end of the reel, it reverses. This method is used in some Concertone machines.

Which one's best? You pay your money and you take your choice. Each system has its commendable features; each has its drawbacks. So shake the dice cup, see how they come out and pick yourself a reversing system.

SOLENOIDS OR BELTS AND PULLEYS

Another feature that you may want to consider is the type of drive mechanism and the number of motors in the machine. Many of today's professional decks use three motors, one for each reel and one for the capstan, along with a system of solenoid controls (see Fig. 3-11). These are electrical controls, rather than the mechanical one found in many of the lower-cost and home-type machines. A mechanical control system usually has just a single motor and a series of many belts, pulleys, slides and other mechanical power-shuttling devices. In a 3-motor solenoid-operated deck, only two belts exist. One is the belt from the capstan drive motor to the capstan pulley; the other is the belt to the index counter. Generally, 3-motor solenoid-controlled decks cost considerably more than the equivalent belt-drive machine. Here, your personal finances and the importance of certain features are the guideposts. If it's worth the extra money to you, fine. But at the same time, keep abreast of new developments in reversible-polarity motors and the auto reversing systems. These features could well make the all-solenoid 3-motor deck too cumbersome for a home instrument. Just make sure you're getting what you pay for.

TRANSISTORS vs TUBES

Chances are, the question of tubes vs transistors won't even figure in your selection of a tape recorder, since virtually all of today's machines use transistors. But if you should run across an older tube-type unit, bear in mind that tubes do have their advantages, not the least of which is their high input impedance. Tubes also have disadvantages, too, but since you'll probably be going all-transistor anyway, we won't belabor the point here.

One feature that you may find very helpful is a headphone monitoring jack. If the tape recorder doesn't have such a jack, it's fairly easy to add one later on, and we'll show you how to do in Chapter 12. Another useful feature is the pause

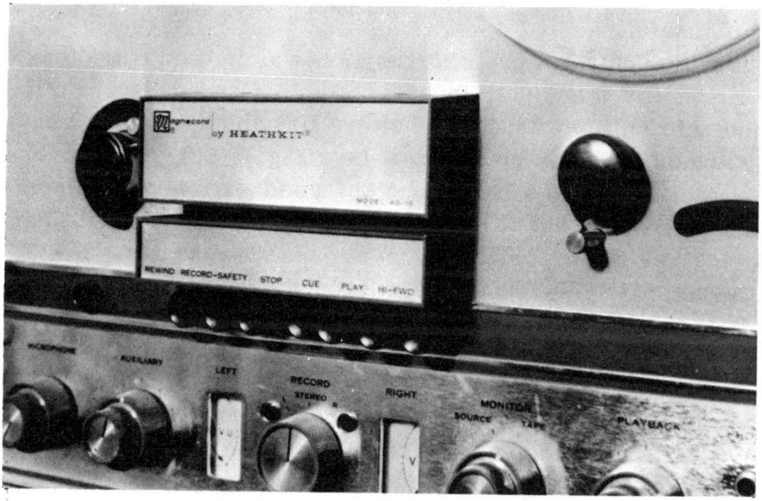

Fig. 3-11. Pictured here are the pushbutton solenoid controls on the all-electric controlled Heathkit AD-16. Solenoid operation is pretty much foolproof and trouble-free as compared to the complex system of belts and pulleys used in mechanically-controlled recorders.

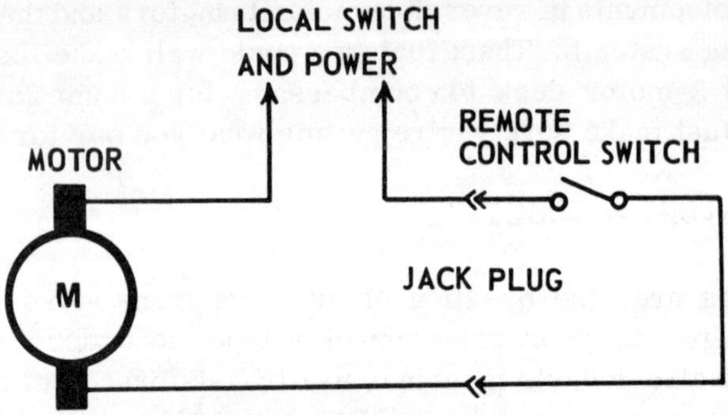

Fig. 3-12. As shown here, a remote control switch is relatively simple to wire in the recorder motor circuit.

control which allows you to set recording levels, monitor, and do just about everything with the recording function turned on but not running; it's simply poised and ready to start in an instant. The pause control can be a spring-loaded lever or it can be a key on a piano-type selector keyboard. It's quite useful.

Remote control is probably the most widely used extra feature on any transistor tape recorder. Chances are you'll find this only on Japanese battery portables, usually as a switch on the microphone itself (see Fig. 3-12). It's useful for dictation, for keeping out unwanted extraneous noises and for getting rid of those crucial pop sounds. In addition to the reel-type portables, most cassette machines have a remote control switch on the microphone.

With this type of machine, a secondary remote switch can be used instead of the one on the microphone any time you want to wire it up. This also goes for foot-pedal controls. Remote is awfully handy for the surreptitious recording of bird calls and other applications where the microphone and recorder are widely separated from the operator.

LEVEL INDICATORS

Recording level indicators are a very important feature. If you set the recording level too high, you'll get overmodulation, resulting in distortion on the tape. There is an ideal area or range in which the recording level should fall. To set the recording level within this range, you must have an indicator of some kind.

There are three major types: the VU meter, the electron-ray tube; and the neon bulb (see Fig. 3-13). The VU meter is most commonly used, especially in transistor tape recorders. The other two types are both effective as level indicators, but are used only with tube-type tape recorders. This is because they require high operating voltages—too high for the transistor tape recorder's power supply. The meter is by far the best indicator for this, so chalk up another benefit of transistorization.

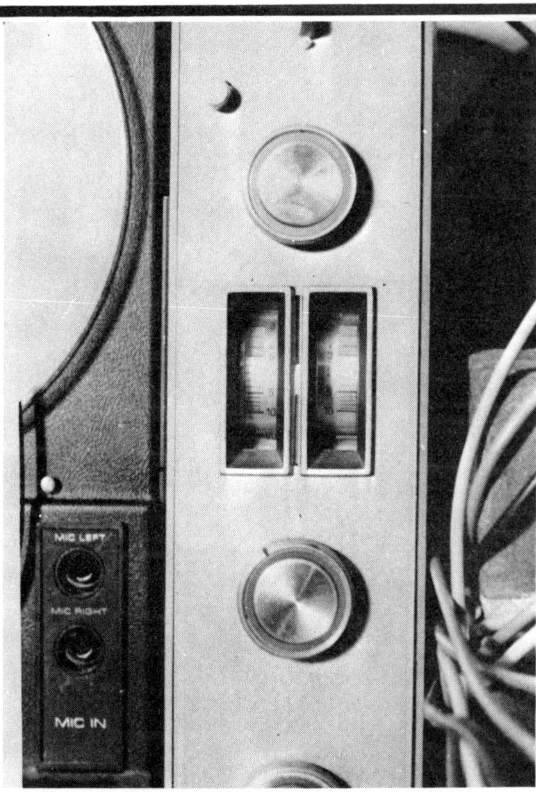

Fig. 3-13. Twin VU meters indicate the recording level on each channel of this stereo tape recorder.

One last thing to look for in a recorder is tape head accessibility for editing purposes. If you want to edit the tape, you have to be able to mark it, cut it, splice it back together and then drop it back into the slot. Some machines let you do this very easily; with others editing is difficult.

Chapter 4

Battery-Operated Recorders

One of the biggest problem areas with battery-operated portable tape recorders is not the machine or the tape but the battery supply itself. Batteries are a problem because too many tape recorder users don't take the necessary few minutes to learn the score—just how well they can expect batteries to perform. As a result, they end up blaming the recorder for poor results or for malfunctions when all the trouble could be cleared up by proper care, feeding, and retirement of the batteries.

The battery has come a long way since our grammar school experiments with oversized dry cells—those big, bulky giants with the little screwknob terminals on top. In fact, batteries are so versatile and so packed with power today that they can do just about any job that's required of them. One of these jobs is driving an electric motor in a tape recorder with unfailing accuracy over long periods of time.

PRIMARY AND SECONDARY BATTERIES

There are two basic battery categories—primary and secondary. A primary type battery is a size D flashlight cell. You plunk down 50¢ for a pair of D cells, slip them into a flashlight, use them until the light gets dim, and then throw them away. These are primary batteries, so-called because they are used once and then must be discarded when they are discharged.

A secondary type battery is the storage battery in your car. When you take power out of it, power must be put back into

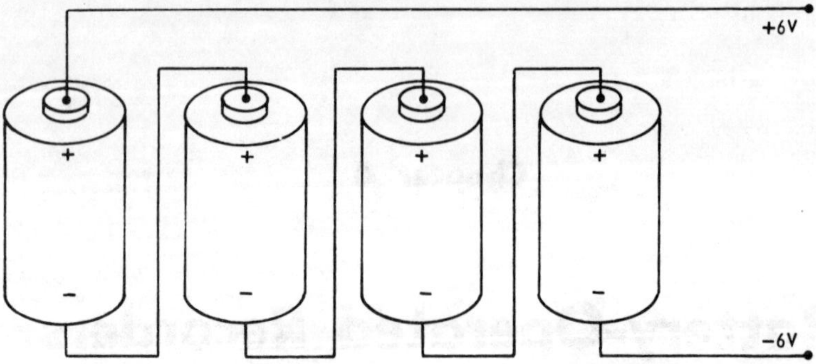

Fig. 4-1. Several batteries or cells can be connected in series for a higher voltage.

it to re-establish its full-charge chemical balance. Secondary batteries aren't just limited to the car; you'll find them in all walks of life today under many guises. The most popular type secondary in widespread consumer use today is the nickel-cadmium cell—a dry-type battery that can be used as a direct replacement for many of the more conventional types. We'll get into this particular battery a little later.

The smallest complete single battery unit is the cell. A cell can be a tiny button-size power plant for a hearing aid or it can be the giant-size dry cells still used to power doorbells in some houses. The basic cell has a nominal voltage of 1.5 volts. To get higher voltages, such cells are connected in series (see Fig. 4-1). Connected this way, the voltages add up. Three cells in series would deliver 4.5 volts; ten would provide 15 volts.

When connected in parallel (see Fig. 4-2), the voltage stays the same but the ampere-hour capacity—the total amount of current available—is increased. The once popular "B" batteries of yesteryear, delivering 45 to 67 1/2 volts, and today's 9-volt transistor radio batteries contain many individual cells stacked or connected in series.

THE MOST COMMON BATTERY

Probably the best known type of battery is the zinc-carbon type, still in common use today in flashlights, transistor radios, and yes, even tape recorders (see Fig. 4-3). In this old-timer, the outer shell is a zinc can which is the negative electrode. The positive electrode is a stick of carbon "sitting" in an electrolyte paste made of ammonium chloride and zinc chloride dissolved in water. The positive electrode is actually manganese dioxide, a coating which forms on the carbon center pole.

Remember those big dry cells in grammar shool, the type used for experiments in elementary electricity? This is a basic Leclanche cell, and the same type cell with minor variations in the amount of electrolyte mix is still used in a wide variety of applications. The battery with "extra bounce" for flashbulbs and the battery for the transistor radio and all the other special purpose varieties are all Leclanche or zinc-carbon batteries with some chemical variation in the mix.

ALKALINE PRIMARIES

Much newer on the scene is the alkaline (or manganese-alkaline) battery which has been the object of massive sales promotion efforts because of its much higher price. If you've noticed this battery, you've also noticed that it's very expensive—anywhere from three to four times the cost of the good old Leclanche type. Yes, it costs more but (and this is a very big but) it lasts a lot longer, too. Typically, a set of alkaline batteries in your tape recorder will last anywhere from four to five times as long as standard zinc-carbon batteries. That's a lot of life. It saves a little bit of money, but it also saves the aggravation of frequent battery

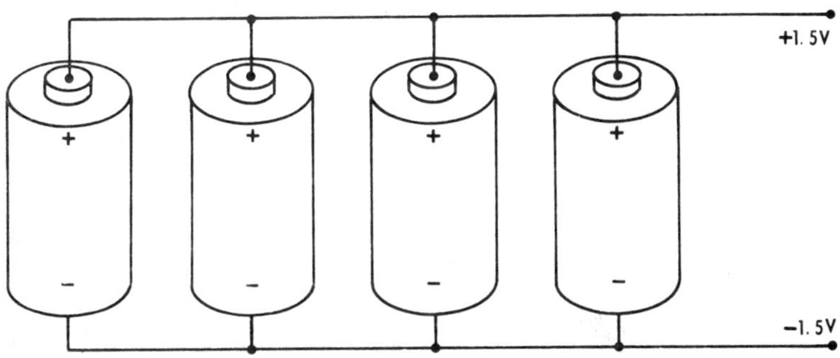

Fig. 4-2. Connected in parallel, the ampere-hour capacity of a group of cells is multiplied but the voltage remains the same as one cell.

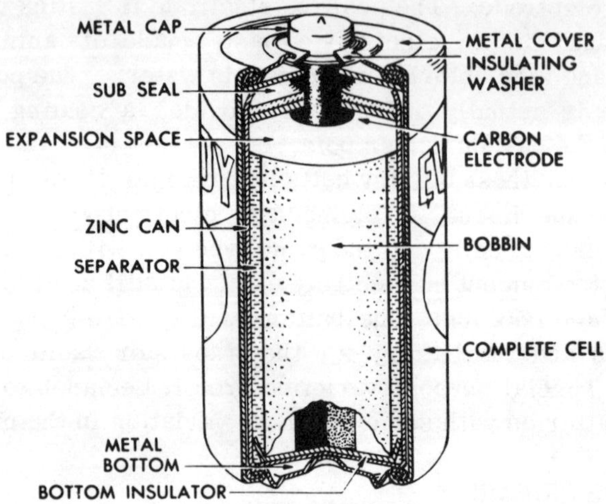

Fig. 4-3. Cross-section of the simple carbon-zinc (Leclanche) cell in its best-known "D" size.

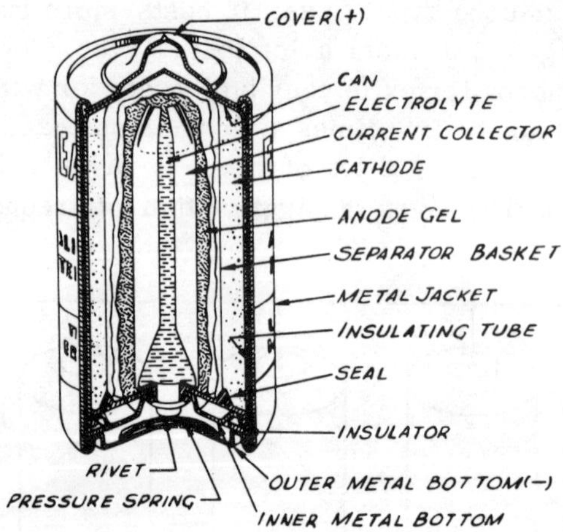

Fig. 4-4. Cutaway of primary type manganese-alkaline battery.

changes and possibly some lost valuable recordings because the old set of batteries had begun to run low.

In appearance, the alkaline battery doesn't look much different from the zinc-carbon—at least on the outside. It has a negative terminal connected to a series of zinc plates—once again zinc is the anode or negative terminal (see Fig. 4-4). The anode is connected electrically to the bottom or outer shell of the can. The cathode or positive terminal is a large area of manganese dioxide, and the electrolyte is calcium hydroxide. The battery is physically and electrically interchangeable with the zinc-carbon battery.

A less familiar variety of alkaline battery is rechargeable. Naturally this particular type is more expensive than the primary alkaline. It's rechargeable, but not indefinitely so. It won't last for a great length of time through thousands of charge/discharge cycles the way that nickel cadmium and lead-acid storage batteries will. It's good for perhaps a hundred or so charge/discharge cycles.

THE MERCURY CELL

The third major type of primary battery is the mercury cell, and this is the one that gives us problems. Its voltage is not the same as the zinc-carbon or the alkaline types which have a nominal voltage of 1.5 volts per cell. So if your tape recorder motor is especially sensitive to voltage changes, you just can't use the mercury battery. But in what areas can you use a battery like this? A mercury battery has a lot going for it once you can choke down the very high initial cost. It has a very long operating life, much longer than the alkaline type. Typically, a mercury battery can last in steady or intermittent service for over a year with good reliability. But, unlike the zinc-carbon and alkaline types, the mercury battery's voltage will drop off very suddenly, once its useful life is near an end. The zinc-carbon and alkaline types drop off gradually in output voltage (see Fig. 4-5).

There are pros and cons in both sides. The gradual drop-off doesn't leave you flatfooted with no batteries at all when you're in a situation that will do with even just a little bit of power. On the other hand, the mercury battery doesn't short-change you on available electrical power. You've got every-

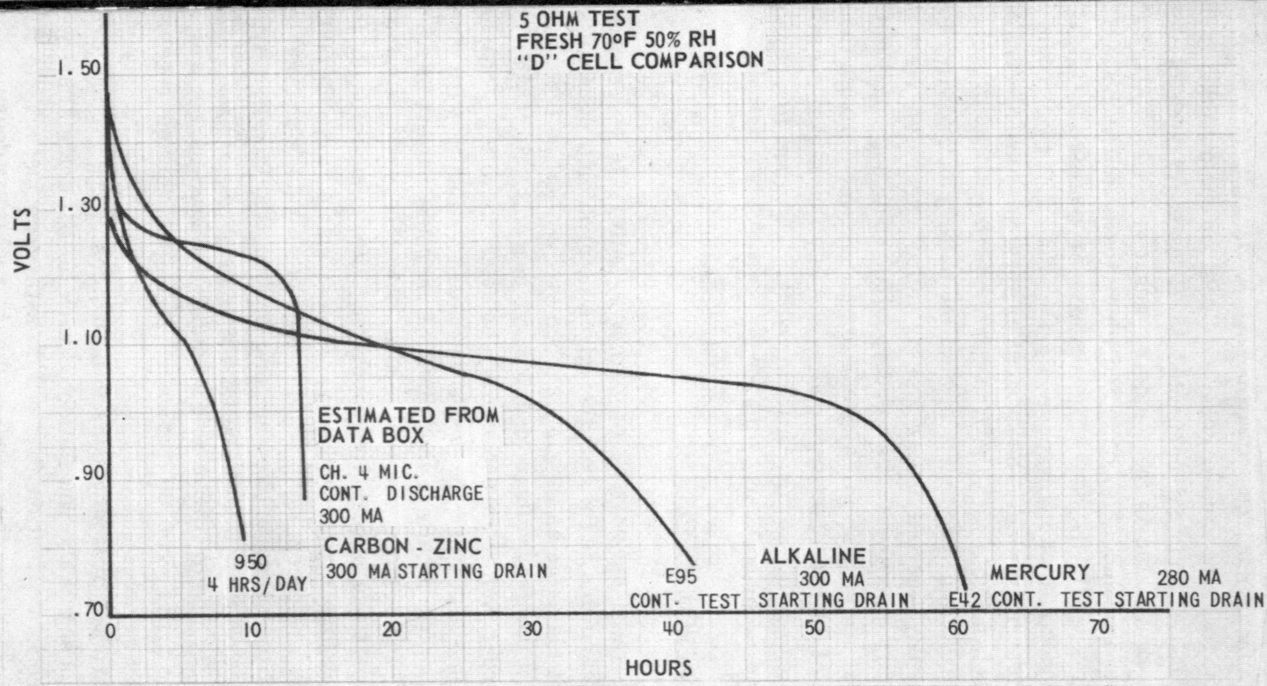

Fig. 4-5. Discharge curves into a 5-ohm load for "D" size cells shows a comparison among carbon-zinc, alkaline, mercury, and nickel-cadmium types. (Graph data supplied by Union Carbide.)

thing it can possibly give you up to the very end, but when it goes, there's no warning. The other types at least give you some indication, because the battery condition meter on your recorder will start to read lower and lower as they deteriorate (see Fig. 4-6).

SHELF LIFE

Another thing to remember when you're buying primary batteries is shelf life. The shelf life refers to the length of time the battery can remain sitting on the shelf, unused, and still retain a useful amount of electrical energy. Remember these are electro-chemical devices, and a chemical reaction continues to go on in the battery even though it's not being used. After a while the chemical activity drops off as the battery becomes consumed, until finally you take down a new, unused battery from the shelf, install it into the tape recorder, and nothing happens.

Generally, you can expect that the shelf life of a battery will extend over a period of one to two years. This means that if you haven't used the battery before a 2-year period is up, you can expect that it will fail very early in service. It's

Fig. 4-6. The tiny VU meter on this Wallensak battery cassette recorder also doubles as a battery condition meter.

much better economy to use the battery intermittently over a long period than to let it sit on the shelf a long time and then try to drain a lot of energy from it all at once. Give the battery a little bit of a chance to recover after each use—it'll reward you for this rest period by coming back with more energy.

A knowledge of shelf life can be of practical use to you in another area—purchasing imported batteries. Advice about imported batteries for tape recorders can be summed up in three words: Don't buy them. The rationale is very simple: A good part of the battery's shelf life has been used up while it's traversing the Pacific Ocean in a ship bottom, then sitting in a California warehouse waiting for trans-shipment to a distribution point somewhere else in the United States, and then sitting on a dealer's shelf. The battery is lucky if it gets there within a short time. Yes, be suspicious. Japanese batteries are fine for transistor radios in most cases and they are all right for flashlights if you remember to take them out after six months to a year to prevent ruinous damage by leakage. But for reliable tape recorder power—forget 'em.

SECONDARY BATTERIES

Now let's take a look at those secondaries that we were talking about. We have the alkaline (alkaline-manganese-zinc) secondary which is not all that popular. There's a very good reason: It's cost represents a substantial investment —considerably higher than the cost of primary types. Also, the number of recharge cycles is very limited. For a somewhat higher cost you can go to the nickel-cadmium secondary battery which has about as many virtues as you could ask for in a rechargeable battery. These batteries are so good that many manufacturers seal them into appliances permanently on the assumption that the battery will not go bad before the appliance itself. The fact of the matter is these sealed nickel-cadmium batteries are so good and so virtually indestructible that many manufacturers give a 5-year guarantee with them. To the best of our knowledge, this is about the only type of battery that has this type of confidence behind it.

The nickel-cadmium battery in its uncharged state has a

positive electrode of nickelous hydroxide and a negative electrode of cadmium hydroxide (see Fig. 4-7). In the charged condition, the positive electrode is nickelic hydroxide and the negative terminal is metallic cadmium. The electrolyte is potassium hydroxide.

One of the unusual features of the nickel-cadmium battery is the way internally-generated gas is handled. Just about

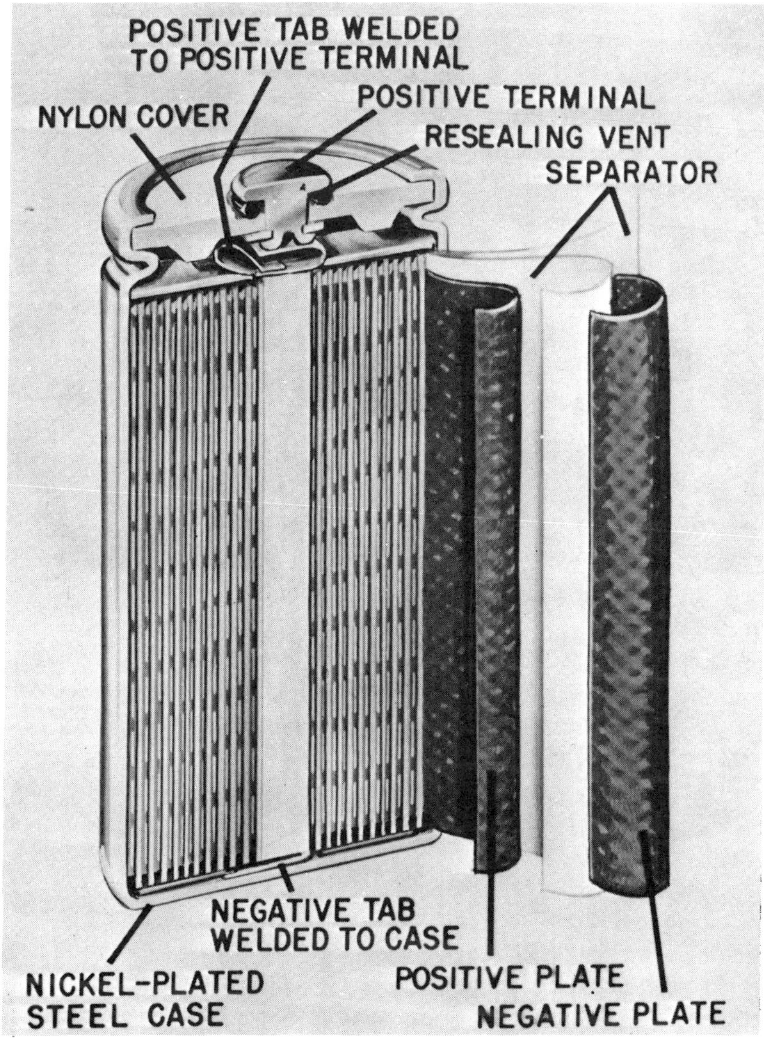

Fig. 4-7. Cross-section of size "D" General Electric nickel-cadmium battery.

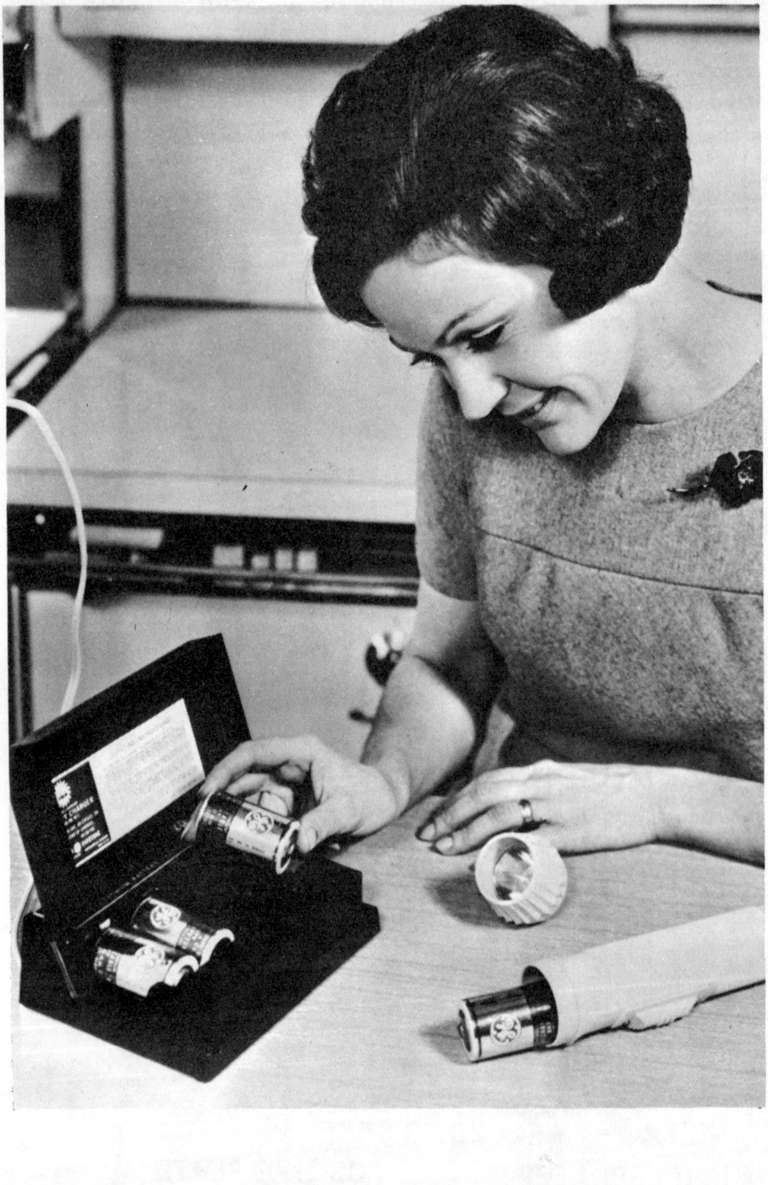

Fig. 4-8. Simple charger by G. E. holds up to four batteries at a time; it is especially good for an overnight nickel-cadmium charge.

any secondary battery generates a certain amount of gas when it's being charged, and the nickel cadmium type is no exception. But what happens to this gas? If you seal the battery, it's possible for gas pressure to build up enough for the battery to explode. In the nickel-cadmium battery, an ingenious solution has been found. The gas (oxygen) simply combines chemically with other materials in the battery to form an inert compound. The results: the battery simply can't be overcharged under any circumstances! Isn't this great for the absent-minded hobbyist who leaves the charger plugged in much too long! And it is equally useful for the photo hobbyist who lets his electronic flash charge longer than necessary.

As with most storage batteries the nickel-cadmium type has one distressing feature: The charge tends to leak off after a time if the batteries are not used. To offset this, it's a good idea to check the battery condition and give it a little charge about once a week or every two weeks if you want to maintain it in topnotch condition and ready to go at a moment's notice. If you're out on a long recording stint with a portable machine, it might also be a good idea to take along a set of very ordinary zinc-carbon batteries in case you exhaust the rechargeable ones and don't have the time to give them an overnight "squirt" with the charger. The old fashioned Leclanches are an excellent standby, and if six months go by without your using them, you can always "plop" them into your portable radio.

By the way, if you have a transistor radio that uses the same size batteries as your portable tape recorder, you can get many extra hours of useful life out of discarded batteries by using them in your radio. They may be unusable for the tape recorder, but not necessarily so for the radio.

RECHARGING PRIMARIES

An emerging school of thought believes that primary batteries can be "recharged" several times after they're exhausted. The idea is that it's possible to reverse some of the chemical action taking place in primary batteries and extend the battery's life by putting it into a charging circuit. There are numerous facts, figures, and tables and other corroborating evidence that show very long total operating

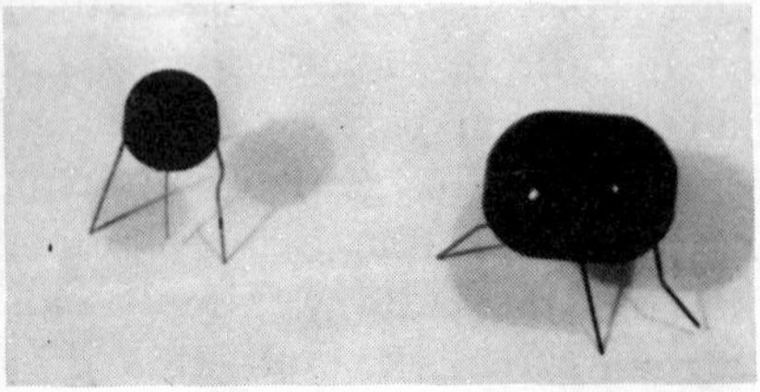

Fig. 4-9. These encapsulated modules from Cordover are designed for do-it-yourselters: (left) battery charger module and (right) a full-wave rectifier bridge circuit.

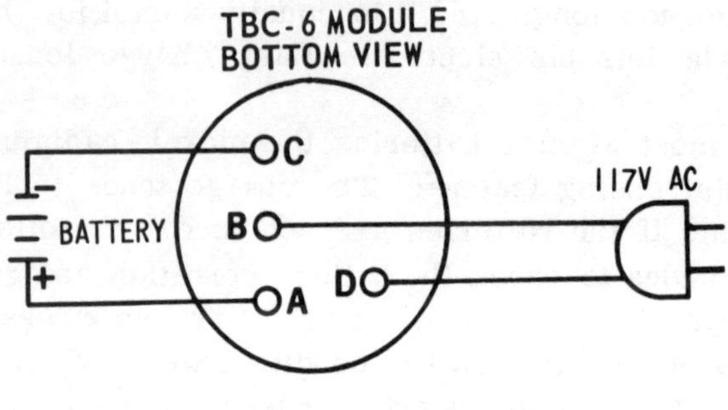

Fig. 4-10. Hooking up the Cordover TBC-6 charger module is a simple matter. Battery or batteries to be charged are connected in series, the negative terminal going to point C and the positive side going to point A.

lifetimes with ordinary non-rechargeable zinc-carbon and alkaline types.

The fact of the matter is that when the battery is dead, it's dead and nothing you do about it is going to change that state. But if you take a battery that has just a little bit of life left, say enough to give a little flicker on a penlight bulb, chances are you can salvage it by an overnight recharge. Don't be very nonchalant about rejuvenating zinc-carbon and alkaline batteries. They don't have the safety factors built into them that the nickel-cadmiums do, and they could rupture or explode if left in the charger too long.

Charging the rechargeable battery can be a simple matter, especially if your tape recorder already has a built-in provision for AC or for an AC adapter. Usually just plugging the recorder into the wall is enough to keep the machinery turning well and to charge the batteries at the same time.

Most battery tape recorders have a switching arrangement of some kind so you can use the VU meter as a battery condition meter. This is an excellent check and could save a lot of trouble.

If you don't pay attention to your battery condition meter, your recordings will tend to get very slow. Voice levels will drop down into a deep baritone and then into a bass. It'll be impossible to understand what's being said, and any high-fidelity recordings that you might be trying to make—well, just be prepared to do them over. And you can't even use the voice recordings that you've made with the batteries in poor condition, because with new batteries the tape will be up to speed again, and the voices will come out sounding like Mickey Mouse.

External chargers are easy to use and provide a simple answer to all battery charging problems. You can buy an inexpensive ready-made one (see Fig. 4-8) or build one yourself (see Fig. 4-9). Most common and safest is the constant-current charger. The circuit for one is shown in Fig. 4-10. Ordinary Keystone battery clips hold the cells, and you can use bypass jumpers if you're not charging four batteries at a time.

Chapter 5

Microphone Basics

Microphones are very funny beasts. They're supposed to do the same job as the human ear, but mechanical equipment is quite different. For one thing, the microphone doesn't have the high-fidelity frequency response of an average adult's ear. The microphone can't sort out the confused babble of many unrelated sounds or voices the way the human ear can. The ear can do this because it's guided by the brain. The mike is guided only by the recording engineer.

MIXED BAG OF TYPES

Microphones come in many shapes and sizes (see Figs. 5-1 and 5-2). Some of them are big bruisers, but most of them today tend to be rather small, including the ones that come with your tape recorder.

Sound waves strike the microphone's diaphragm, a semi-flexible disc that works very much like the eardrum, and make it vibrate. The number of vibrations per second corresponds to the frequency of the sound wave striking the instrument. The diaphragm is linked mechanically to the transducing element, the electro-mechanical part of the microphone that translates mechanical motion into electrical energy.

There are basically four kinds of transducing elements used for tape recording. These include piezo elements (rochelle salt crystal or barium titanate ceramic); dynamic (or moving coil) type; velocity (or ribbon) type; and condenser type.

The piezo type depends on the unique ability of the crystal or ceramic slab to generate electricity whenever it is stressed mechanically (see Fig. 5-3). This same type of element is used as the pickup in low-cost record players; the needle riding in the record groove vibrates against the piezo element. Piezo-type microphones are very inexpensive and frequently accompany tape recorders as original equipment (see Fig. 5-4). Trouble with a crystal or ceramic microphone is that the frequency response is very limited, usually cutting off at around 8,000 Hertz for the ceramic, and lower for the crystal.

The dynamic microphone generates an electrical current by vibrating a coil of wire in the field of a permanent magnet (see Fig. 5-5). The resulting signal is of much better

Fig. 5-1. This classic-shaped RCA ribbon microphone with adjustable pickup patterns is still used today in many radio stations. The ribbon mike has characteristically round, warm, natural sound, and is an excellent high-fidelity instrument.

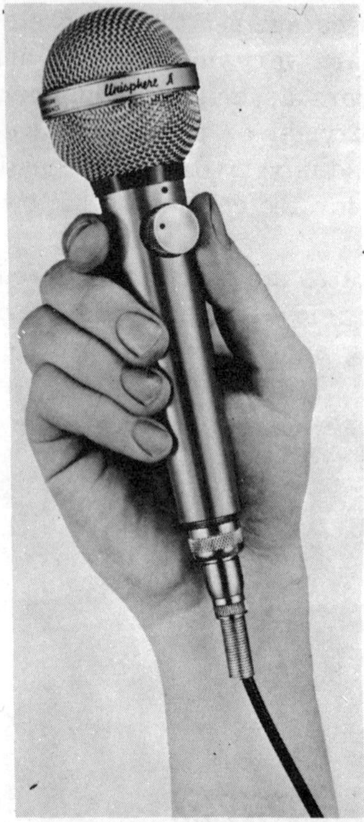

A

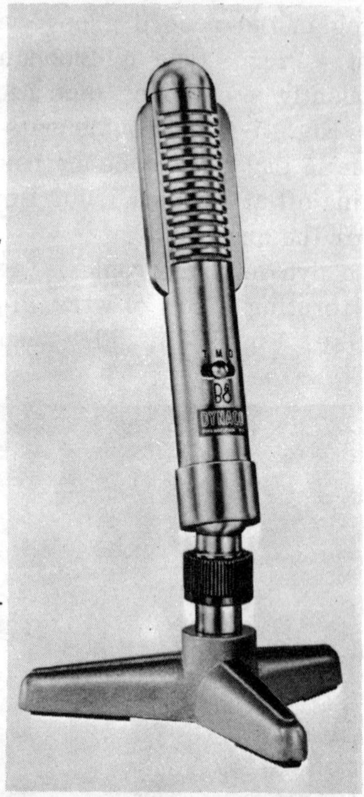

B

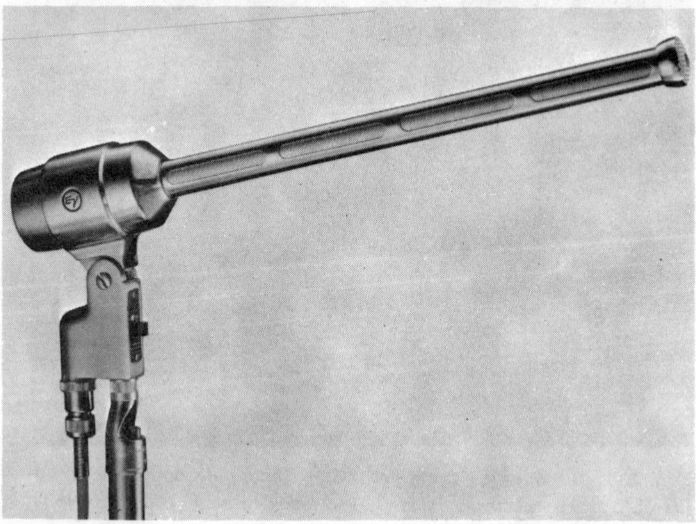

C

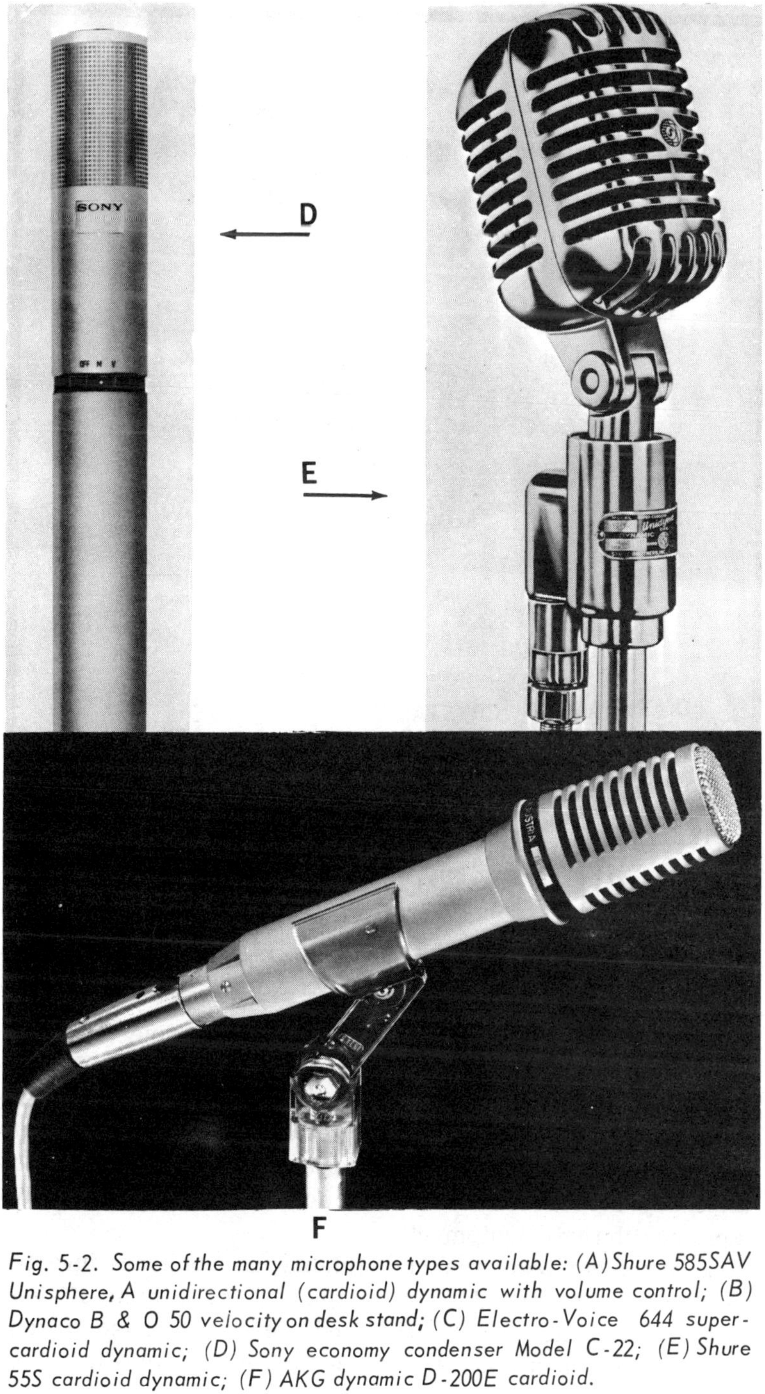

F

Fig. 5-2. Some of the many microphone types available: (A) Shure 585SAV Unisphere, A unidirectional (cardioid) dynamic with volume control; (B) Dynaco B & O 50 velocity on desk stand; (C) Electro-Voice 644 super-cardioid dynamic; (D) Sony economy condenser Model C-22; (E) Shure 55S cardioid dynamic; (F) AKG dynamic D-200E cardioid.

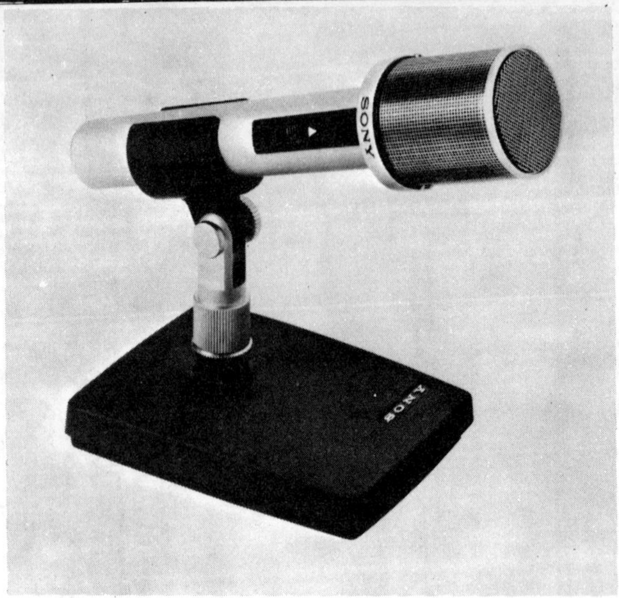

Fig. 5-2G. Sony F-121 cardioid dynamic.

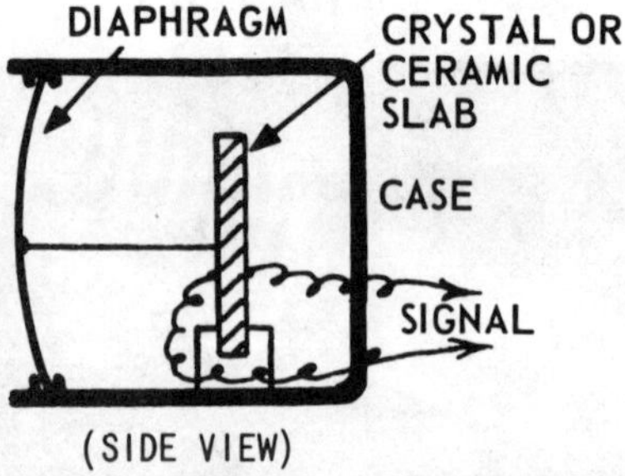

Fig. 5-3. Sketch of the "insides" of the simplest type—the crystal or ceramic. Sound waves striking the diaphragm cause the crystal or ceramic slab to vibrate, generating a small electric current.

fidelity than is possible with a piezo microphone (see Fig. 5-7). As a result, the dynamic has gained a great deal of acceptance both in home recording and in the broadcasting industry. The dynamic has gone through many improvements and downward price revisions in the last few years, and today many tape recorder manufacturers offer low-cost dynamics with their equipment.

One of the earliest true high-fidelity microphones invented was the velocity or ribbon microphone. Many radio stations still use their aging velocities and not just because of sentimental attachment to that 30-year old mike (see Fig. 5-8).

Fig. 5-4. Typical ceramic microphone, the Electro-Voice Model 715.

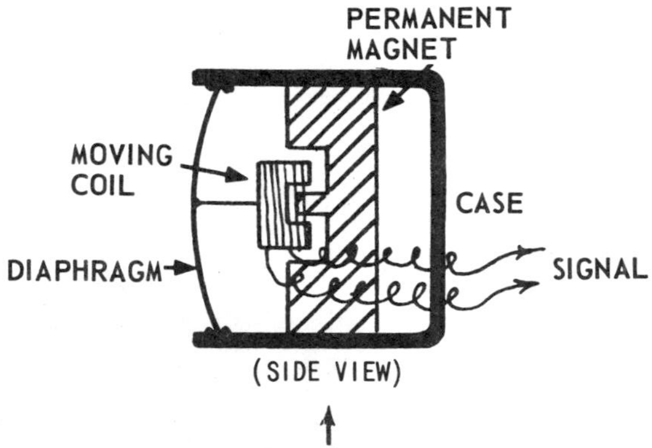

Fig. 5-5. A dynamic microphone has a coil of wire attached to the diaphragm. As this coil vibrates in the field of a permanent magnet, a small signal voltage is generated.

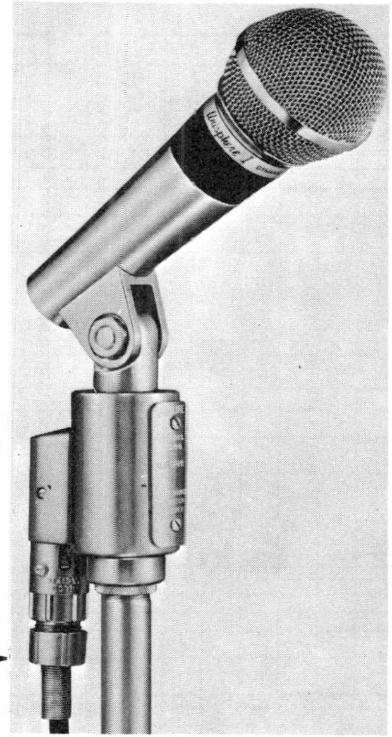

Fig. 5-6. A typical dynamic microphone, the Shure 566 is a cardioid with a built-in shock mount and wind and pop filters.

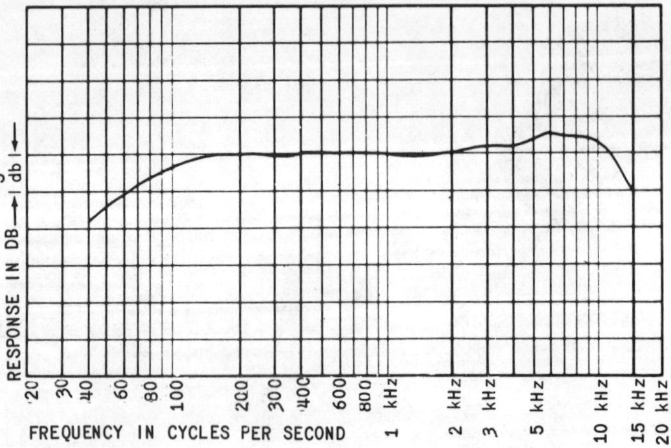

Fig. 5-7. Frequency response curve for Electro-Voice 635A, an omnidirectional dynamic mike that's very popular among performers for hand-held use.

Fig. 5-8. An ancient velocity microphone still in use in a well-known radio station.

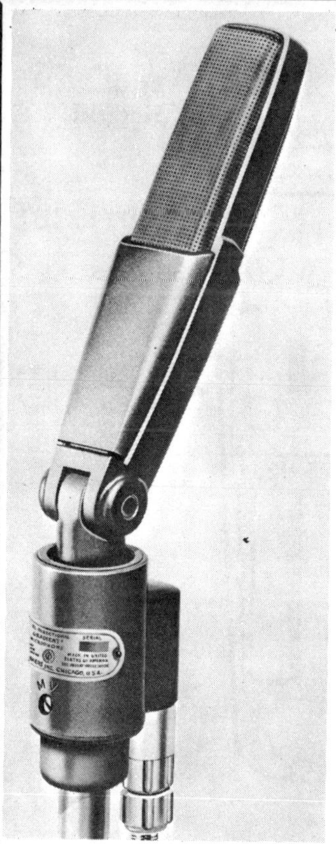

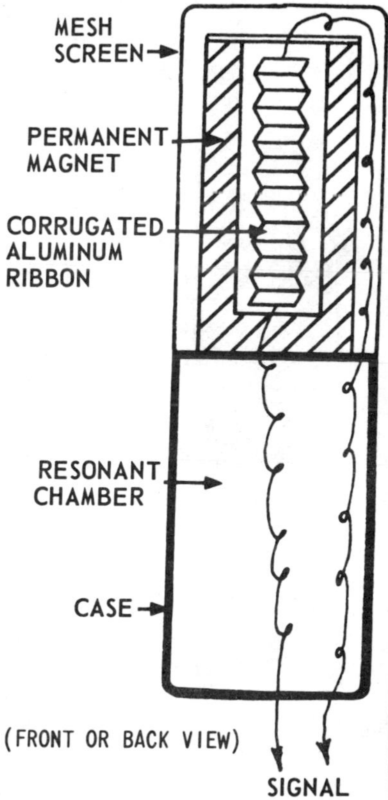

Fig. 5-9. Modern velocity microphone by Shure (Model 300) has silky-smooth frequency response; it is small, lightweight and relatively inexpensive.

Fig. 5-10. This cross-section of velocity microphone shows the lack of a diaphragm. Sound waves strike the fragile, corrugated aluminum ribbon, making it vibrate between the poles of an elongated permanent magnet. This vibration causes a signal current to flow in the aluminum.

Even old velocities continue to work well, providing exceptionally smooth frequency response (see Fig. 5-9).

The microphone in Fig. 5-10 has no diaphragm. Instead, the sound waves strike a corrugated, very frail, aluminum ribbon causing it to vibrate between the poles of a very powerful permanent magnet. The ribbon generates a feeble electric current, and that's all there is to it. The velocity microphone is known to be a very fragile instrument, although they

Fig. 5-11. This Sony condenser microphone, Model C37, has adjustable pickup patterns.

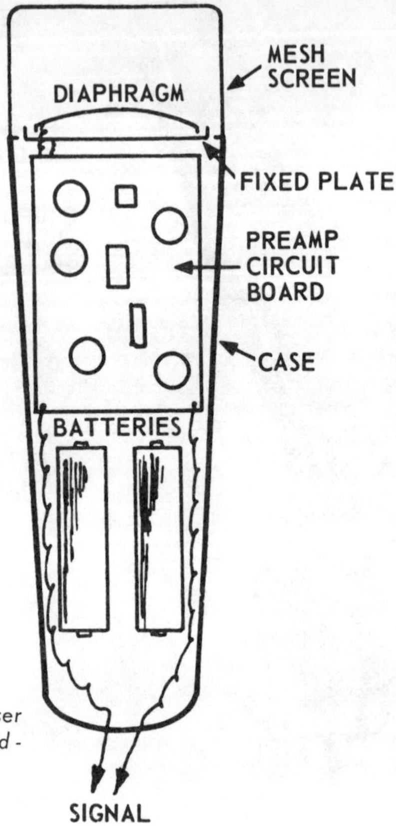

Fig. 5-12. This inside view of condenser microphone shows the vibrating condenser plate, FET circuit board, and battery power supply. Mikes of this type have superb frequency response.

are usually damaged only when wind passes through the case and wrecks the ribbon element. For this reason, it can't be used outdoors and should be kept off fast moving dollies of the type used in TV studios.

ENTER THE CONDENSER

The most popular microphone for true high-fidelity recording is the condenser type (see Fig. 5-11). Like the velocity, the condenser has been around for a long time, but its true value has been recognized only within the last decade or so. Back in 1932, there wasn't much use for this mike's usable response—up to 10,000 Hertz then. Recordings and radio broadcasts just couldn't handle such a wide frequency response.

In the halcyon days of high-fidelity recording just before stereo came on the scene, when the single-microphone technique had been developed to a fine art, it was the condenser microphone hanging high above the head of the orchestra's conductor that made this technique possible. This microphone is just as useful in pairs for stereo recordings. It's an extremely versatile and very expensive instrument.

In the condenser mike, a diaphragm is the active element in the electrical circuit (see Fig. 5-12). It actually forms one plate of a capacitor, and as sound vibrations strike the diaphragm its capacitance changes in accordance with the varying distance between the plates. This changing capacitance varies the resonant frequency of a special preamplifier circuit, and the preamplifier produces an output signal voltage at appropriate audio levels.

Until fairly recently, rather hefty power supplies and preamps were needed for these microphones, and it was quite a neat trick if you could operate two mikes from the same power supply. However, transistor technology has come to the rescue, and some of the latest condenser microphones have the entire preamplifier and power supply within the mike's case.

What about prices? Well, the crystal and ceramic types cost anywhere from $2 to $10 each. Dynamic mikes can run from $5 to a couple of hundred dollars; figure $50 average for a good, general purpose mike. Velocity microphones cost from $50 to $200; figure $50 to $100 each for good quality home-type microphones. Condenser microphones run from about $125 to $400 each. There's no such thing as an average mike there. The cheapest one will give far better quality than you're probably used to.

DECIPHERING THE SPEC SHEET

When you start looking around at microphones, you'll discover that there are many difficult-to-understand features.

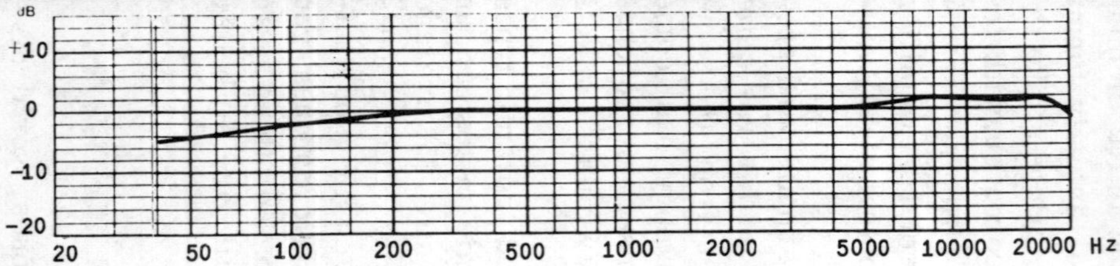

Fig. 5-13. Look at this almost ruler-flat frequency response curve of a condenser mike—the Neumann KM84i.

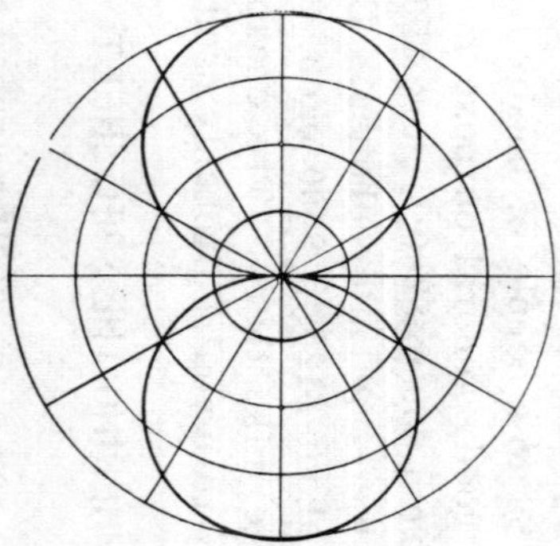

Fig. 5-15. Here is an "idealized" polar pattern for bidirectional velocity and condenser microphones. This figure-8 pattern is the natural one for ribbon mikes, and it is induced electronically in some condenser types.

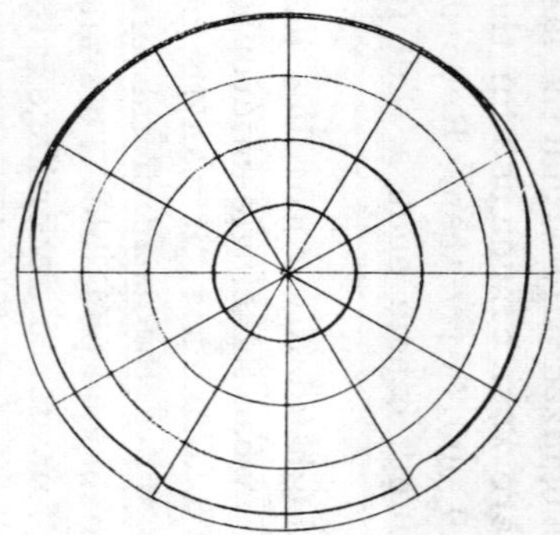

Fig. 5-14. This omnidirectional polar pattern shows how a microphone picks up almost equally well from all directions, no matter how it's "aimed."

A spec sheet will give you a frequency response curve that looks very interesting and very incomprehensible. If it has a sharp knee at the high-frequency end (see Fig. 5-13), it may be due to an artificially-created resonance point to help raise the mike's high-frequency response. This is a perfectly legitimate trick and is used on some of the best microphones. It gives them a little bit of extra brilliance at the high end.

Typically, many dynamic microphones have a usable flat response up to only about 12,000 Hertz. This is fine for the average set of human ears, since most of us hear only up to about 12,000 Hertz anyway, notwithstanding many high-fidelity claims to the contrary. But you really do need to capture those higher frequencies, since they contain many rich harmonics that give the music its particular timbre and quality.

With a little shopping around, you can find a dynamic mike that's reasonably flat up to 15,000 Hertz. A dynamic type with usable response to 20,000 Hertz just doesn't exist. If the spec sheet says so, don't trust the manufacturer. We've seen this kind of frequency response claim on dynamic mikes tagged at $8 to $10! Who's kidding who?

For general purpose home recording, the dynamic is probable the best choice. It's very rugged as compared with other microphone types and will take an incredible amount of abuse. We've seen dynamic microphones used as hammers to drive nails into wood blocks and dunked in a bucket of water. The mikes continued to produce just as if nothing had happened. Try the water routine with a crystal microphone and goodbye mike; try driving nails with a velocity or condenser type and you'll be shopping for a new one.

PICKUP PATTERNS

There's another characteristic that's very important—the pickup pattern. Most microphones are by nature omnidirectional; they pick up sound equally well from all directions (see Fig. 5-14). This is true for most mikes except for the velocity, which has a bidirectional pattern—it picks up with equal sensitivity from two directions, directly opposite each other, but it is completely dead at the sides (see Fig. 5-15). This pickup pattern, which looks something like a figure 8,

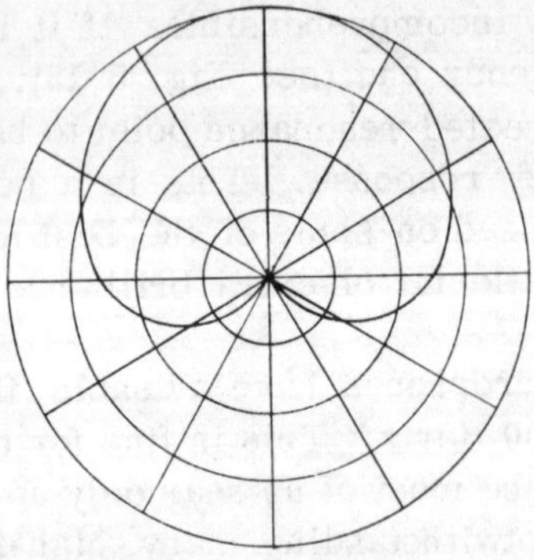

Fig. 5-16. A cardioid (heart-shaped) polar pattern for unidirectional microphones. Notice the mike is "dead" at the rear, but has some side-lobe pickup.

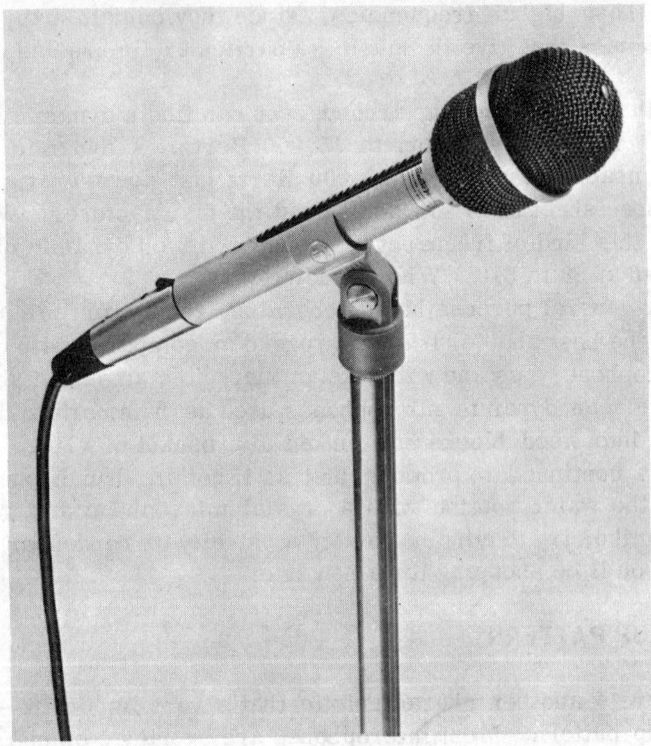

Fig. 5-17 Ribs on the back of this Electro-Voice cardioid microphone let in just enough sound from the rear to cancel any rear pickup. It's this combination of holes and vents that make a cardioid mike out of what is naturally an omnidirectional type.

can be especially attractive for recording duets on a single microphone or for picking up reverberation from a wall while getting the main sound of an instrument from the front of the mike. Other than that, the velocity microphone's dual pickup pattern can be a nuisance. Any attempt to use it for a musical instrument duet can cause phase cancellation and other distortion problems.

THE MOST USEFUL PATTERN

An extremely useful pickup pattern, generated artificially by changing the microphone's case design, is the cardioid (see Fig. 5-16). As you can see, it is a heart-shaped pattern which is roughly unidirectional. The cardioid is most sensitive directly in front, completely dead directly behind, with a certain amount of sensitivity at the sides (see Fig. 5-17). One use for this type of pattern is eliminating unwanted background noise. Orient the microphone with its back towards an unwanted noise source and most of the noise will be cancelled. The selective directionality of the cardioid lends itself well to stereo recordings. A matched pair of cardioid mikes will go a long way toward making stereo recordings come alive in your home.

There's only one trouble with the cardioid pickup pattern: In the process of achieving this pattern from a naturally omnidirectional microphone, the high-frequency response is degraded. If we start with an omni mike that has a 15,000 Hertz response, the cardioid version may have a useful frequency response only up to perhaps 12,000 Hertz. If you want a cardioid flat to 15,000 Hertz, then you'll have to start with a basic element that has a much higher omnidirectional response. This is a very touchy order to fill with a dynamic mike, and there are very few cardioid dynamics with a truly flat response up to 15,000 Hertz, especially at a medium price ($50 or so).

The pickup pattern for the condenser microphone, by the way, can be varied depending on the model you purchase. Most are available in either omni or cardioid patterns, while some have a system of movable or adjustable parts so you change the pickup pattern to suit your needs.

GETTING THE RIGHT IMPEDANCE

Impedance-matching is another crucial factor. Your tape recorder has a certain input impedance. Generally, it is

Fig. 5-18. With a screwdriver you can adjust the impedance on this Shure ribbon mike.

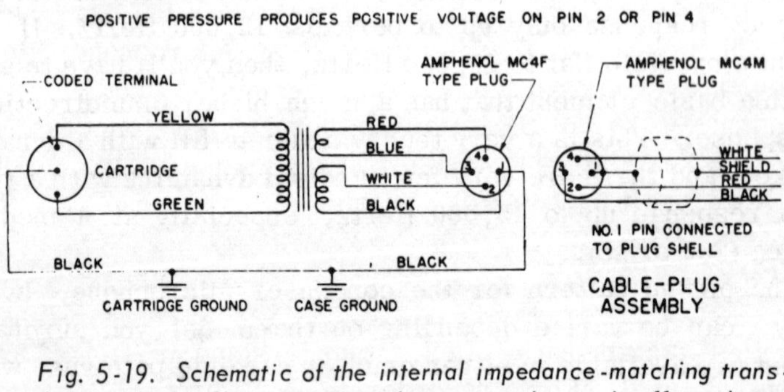

Fig. 5-19. Schematic of the internal impedance-matching transformer in the Shure 565 Unisphere microphone. It offers choice of unbalanced (high-impedance) output or balanced (low-impedance), depending on the choice of wire connections used. The most common is unbalanced (for home tape recorders), which uses the shield braid and the red wire.

high—from 10,000 t0 100,000 ohms, especially if it's a tube-type tape recorder. If it's a transistor-type tape recorder, it may or may not have a high input impedance, depending on what the manufacturer did in the final design. It may also have a low impedance—about 1,000 ohms. It's important that you know this figure, since any microphone that you use should have its impedance matched to that of the tape recorder. Crystal and ceramic mikes are always high-impedance types. Dynamic and ribbon microphones characteristically have low impedances—50 to 100 ohms.

If you want to use a low-impedance mike with a high-impedance input, a matching transformer is needed. Some manufacturers include a matching transformer right in the dynamic microphone's case. It may simply be a matter of a screwdriver slot adjustment to select the proper impedance (see Fig. 5-18), or it may be a question of changing the wire lead connections at the plug (see Fig. 5-19).

In some cases you may have no option at all, but instead must add an impedance-matching transformer in the line (see Fig. 5-20). This transformer is a hidden cost. You may decide to buy a $30 "bargain" dynamic microphone, then find you need an impedance-matching transformer for another $10. Your bargain is no longer a bargain. The transformer must be considered part of the entire package cost.

Even the so-called low-impedance transistor tape recorders often need some kind of impedance-matching transformer, since there is still a great mismatch between the machine's 1,000 ohms and the 50 to 100 ohms at the microphone. Here, a microphone-matching transformer can't be used since the impedances are not the same as those usually encountered. Instead, a miniature transistor-type transformer, costing 89¢ or so, will probably do the trick.

OUTPUT LEVEL

Another factor that's often misunderstood is signal level or output level. For a recorder to work properly, a signal voltage somewhere in the range of 0.5 volt to about 2 volts should be available. But don't look for this kind of information on the microphone spec sheet; it isn't there. Instead, it will say something like, "signal output -57 db at high impedance; -59 db at low impedance." Translated into usable

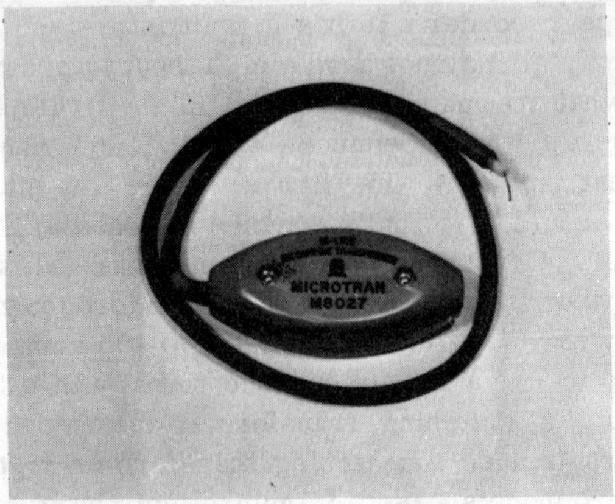

Fig. 5-20. The Microtran in-line impedance matching transformer sells for about $9. It attaches at end of a low-impedance mike cable to provide unbalanced, high-impedance output.

English, the figure -57 db represents a higher signal voltage than the -59 db. The figure actually is taken with respect to a reference signal voltage generated by a specific sound level. Trouble is, every manufacturer has a different way of measuring this.

As a general rule of thumb, signal levels from -55 to -58 db will work well with most tape recorders. If you get down to a low signal level like -60 db, you're in trouble and will need some kind of preamplification between the microphone and the tape recorder. You'll get low signal outputs like this with velocity microphones and with a very few dynamic microphones. Crvstal and ceramic microphones have extremely high signal outputs, and if anything, they'll wreck your recording level meter by "pegging" it too hard.

WHAT ABOUT THOSE WIRES?

When you buy a microphone, you may be dismayed to discover that it has no plug or connector of any kind on it, just some bare wires coming out of the end of the cable. There's a reason for this. The microphone manufacturer just has no way of knowing what kind of equipment the microphone is going to be used with. Different kinds of equipment take different types of connectors.

If you have the usual tape recorder, the microphone will require a standard phone plug, 2-conductor plug that somewhat resembles the plug used on a telephone switchboard. Plug types are detailed in Chapter 12. But, you ask, there are four wires coming out of that cable. Which ones do I connect to the plug and how?

This is a very legitimate question, and it depends on the microphone. If you have a low-impedance dynamic type, you may want to add a line impedance-matching transformer. If such is the case, you'll use two or three of the four wires coming out of the case.

One of these wires is a shield, a braided wire that surrounds the other wires. The shield is usually connected to the microphone case and it serves as a common or ground connection. If there are three other wires coming out of the cable, the black one most likely will be another ground connection, connected to the case inside the microphone. You can probably twist this together with the shield. The two remaining wires will come directly from the coil. This is called a "balanced output" and may be referred to as such in the spec sheet. This simply means that neither side of the microphone coil is connected to ground. One of the first things you'll do is change this into an unbalanced output by connecting one of these wires to ground.

If you're using two identical microphones—a matched pair—be sure that the same color wire from the coil goes to ground in both microphones. This will prevent any phase distortion or phase cancellation when you make a recording.

You may have a transformer already built into the microphone which enables you to choose either a high- or low-impedance output simply by selecting certain pairs of wires coming out of the cable. The spec sheet will tell you this either in the form of a schematic drawing or simply by spelling out which colors are which impedances. If this is the case, just snip off the unused wires and forget about them. The transformer output will be unbalanced.

One thing you've probably realized by this time is that you'll have to learn to be a littly handy with the soldering iron, if you aren't already. Setting up microphones and doing a good quality tape recording at home isn't a ready-made mail-order house operation, unless you want to spend extra money to have someone else set it up for you. So warm up that soldering iron, roll up your sleeves and let's go!

Chapter 6

Getting the Most From Your Mic

Buying the right microphone is only half the battle. If you don't know how to set it up and use it properly, you're still not going to get the best possible results out of your taping session. Professional audio engineers sometimes spend years learning the tricks and techniques of their trade.

While you may never reach the pro's degree of perfectionism, there's no reason why you can't start to get professional-sounding recordings at home right away. To be sure, there's a certain amount of trial and error involved the first few times around. Even after you've become pretty proficient in microphone placement, you'll run into knotty problems that will force you to resort to the trial and error bit again.

KNOW YOUR GEAR

The first step in learning to use a microphone properly is to know as much about your equipment as possible. What is the actual usable frequency response of your microphone? What shape is its polar pickup pattern? What is its nominal impedance? What special characteristics does it have that might help you? The answers to these questions, plus another group of questions that you'll have to ask yourself at each recording session, spell out the basic ground rules for microphone placement.

Let's take frequency response: Certain types of musical recordings or microphone pickups require better frequency response than others. Take for example, a singer who accompanies himself on a guitar. If he is singing solo, you can use one microphone to record his voice and the other to pick up the guitar.

Fig. 6-1. The two-microphone setup here, for a singer with guitar accompaniment, shows the need for optimizing microphone characteristics. The very wide-range Shure velocity is used for guitar pickup; the omnidirectional dynamic with wind screen and pop filter (Electro-Voice) is used for the vocalist. The "omni" pattern helps prevent signal fading when the singer turns her head during the performance.

If you have two microphones that are not matched, one may have better frequency response than the other. You will probably want to use the wider-range mike on the guitar because of the instrument's very rich harmonics. The human voice will reproduce well on an instrument with a somewhat less-extended frequency response. If one mike has a usable response up to 12,000 Hertz and the other goes up to 15,000 Hertz, with all other things equal, you'll use the better mike at the guitar (see Fig. 6-1).

POLAR PATTERN

Next thing to consider is the polar pickup pattern—the pickup characteristics of the microphones. In Chapter 5 we talked about omnidirectional, bidirectional, and cardioid pickup patterns. If you have a choice of microphones to use with our guitar-playing singer, this is where we have to be very careful about pickup pattern. You may want to "close mike"—

use the microphone very close to the **vocalist's** mouth for a particular effect (see Fig. 6-2). "Close miking" can be very helpful, especially if there's a danger of the voice being drowned out by the sound of the guitar or other ambient noise or musical material. "Close miking" is needed also if the singer's voice is just naturally very low. We can make a whisper into a roar with this technique.

But there's one big danger in "close miking"—you can't use a cardioid or a bidirectional microphone. You must use an omnidirectional type. The reason: When you "close mike" with a highly directional microphone, the slightest head motion off-axis will cause a drop anywhere from 2 to 5 db in the recorded level. You'll get a kind of wavering volume level which is very distressing and unprofessional.

If "close miking" isn't absolutely essential, and if your singer "bobs" his head quite a bit, you might try using a lavalier microphone as they often do on television (see Fig. 6-3). This is a small lightweight microphone suspended by a cord around the performer's neck. The advantage here is

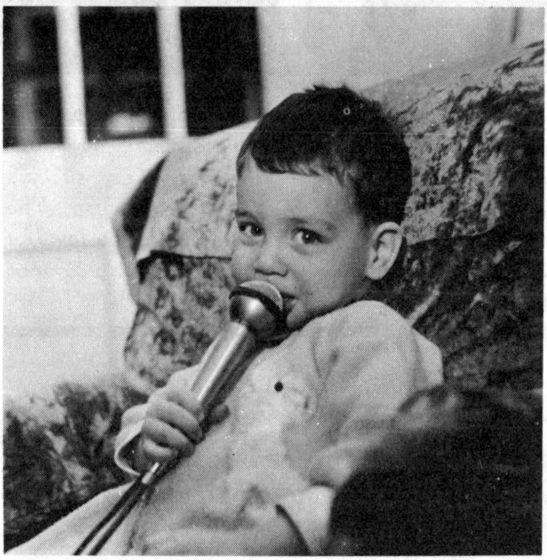

Fig. 6-2. "Close miking" helps eliminate a lot of background noise, since very low recording level setting can be used at the recorder. It also picks up very low voices on mike.

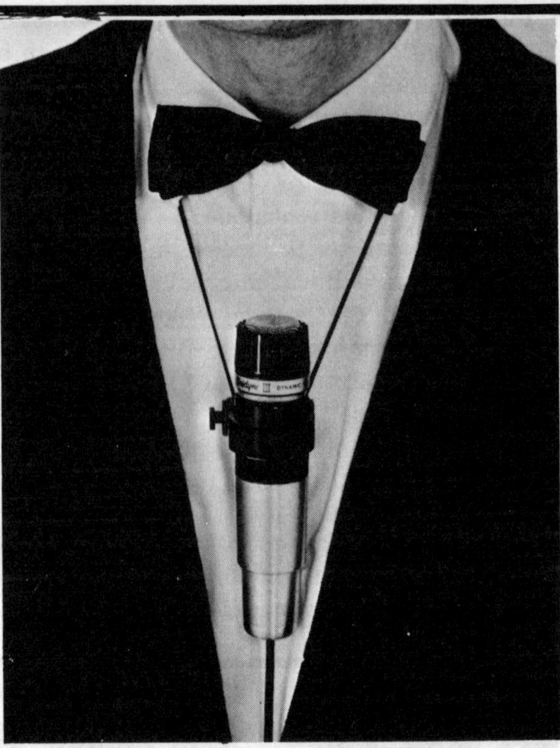

Fig. 6-3. Hanging a mike on a lavalier keeps it the same distance from the performer's mouth, no matter what he's doing. Small, lightweight mikes, such as this Shure Unidyne III are a must for this use. The frequency response falls off somewhat in the lavalier position, so it's best for speakers, not quite so good for singers.

that the microphone remains at a fixed distance from the performer's lips. Trouble is, lavalier microphones don't always have very good frequency response and you may have to be satisfied with a pickup that cuts off at about 10,000 Hertz.

WINDSCREENS

While we're on the subject of the guitar-playing vocalist, if you have a choice of two microphones (and we speak here from bitter experience) consider this possible situation. On hand, one very wide-range (to 15,000 Hertz) omnidirectional microphone and on the other a well-defined cardioid type with a response up to 12,000 Hertz. The omni's response is so superior that you're tempted to use it for the guitar and use the somewhat lower-frequency cardioid for the voice. But this arrangement is ruled out by the "close miking" that's necessary, since the cardioid pattern is so directional that head-bobbing makes the recorded sound level go all over the place.

Another thing: The omni also has a triple wind screen and "pop" filter—very advantageous when close miking. Without the windscreen/pop filter, the slightest explosive consonant sound voiced by the vocalist will come across like so much thunder. In this case, frequency response be damned. Reverse the mikes and use the wide-range omni for the vocals and the cardioid for the guitar.

PROPER PHASING

Another important factor in microphone placement is phasing. Let's assume that you've connected the microphones for proper phasing as discussed in Chapter 5. But, there's another problem, and this occurs with one particular microphone type—the velocity (ribbon) microphone. You'll recall that the velocity mike has no diaphragm, but instead has two sensitive surfaces where sound waves can enter and strike the fragile corrugated aluminum ribbon. Sound waves coming from side A are exactly 180° out of phase with sound waves coming in from side B. If we place two vocalists singing a duet on opposite sides of the microphone (unless one is a soprano and one is a tenor), we'll get a certain amount of phase cancellation and phase distortion. The moral of the story is don't put them on opposite sides—put them both in front (see Fig. 6-4).

The ribbon microphone really is a standout performer when used properly. It's finest hour comes when it's used for recording musical instruments. Probably the best obtainable recording of a grand piano can be made with the ribbon microphone by angling the mike about 10 feet from the piano with its opposing face pointing directly away from the piano to pick up room reverberation (see Fig. 6-5). This will work well in a room that's acoustically alive or in a large recording studio or concert hall.

Fig. 6-4. The pros set up velocities to pick up room reverberation from the side opposite the sound source. Here, in a Capitol Records recording session, Leonard Pennario is the pianist, Vladimir Golschmann conducting.

If you work in a studio that's covered with sound-absorbing material, there's not going to be much sound reflection. You're much better off in an acoustically tuned room such as a concert hall or auditorium. In the living room, you may have to experiment with various setups. The piano is a difficult instrument to record at best, and the living room is not the best possible place to do it. You may want to set up a velocity microphone some distance away to pick up sound waves from a distance, along with a very rugged (dynamic) microphone buried in the maw of the instrument (see Fig.

6-6), and then mix the output of the two. You might want to try velocity microphones at different distances from the piano to provide a slight coloration and difference in phase-arrival time. This will give a richer recorded sound, especially if the final mix is on the same channel.

POST-SESSION MIXING

If you're recording a single instrument with two separate microphones, the best bet is to record them as stereo signals and then mix them electronically on another tape. This way you can control the exact amount of mix from each channel on your "dub" and experiment for the best possible effect without altering your original take. If you don't like the results, you can always do it over without disturbing your recording "master." Doing this requires two tape recorders, of course. This is basically the technique used in profes-

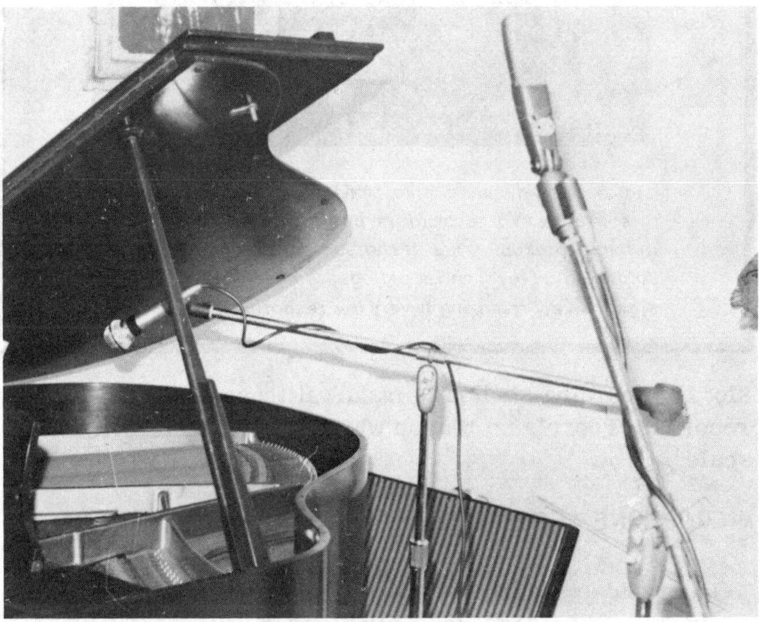

Fig. 6-5. In this living room setup, two mikes are used and mixed at the recorder. One mike, a Shure Unisphere cardioid, is placed close to the piano sounding board. The second mike, a Shure 300 velocity, provides brilliance and room reverberation pickup from its rear.

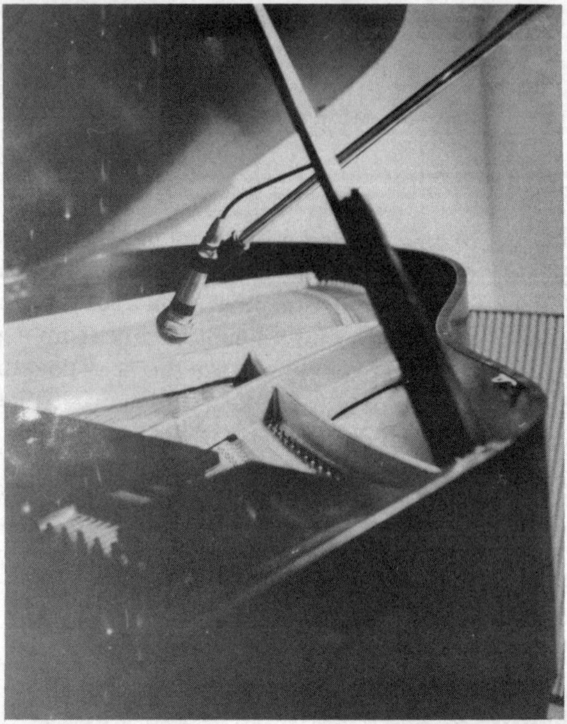

Fig. 6-6. A dynamic mike can take quite a beating, so this Electro-Voice omni is buried in maw of piano for intimate pickup. This technique is also an aid in reducing noise level on recordings, since it provides high signal level, requiring a very low recording level setting.

sional recording studios to make all those luscious-sounding records. There's no reason why you can't do it on a smaller scale.

MICROPHONE ACCESSORIES

Once you've learned some of the fundamentals of microphone placement and use, you'll start getting into the realm of accessories. A very basic item is the stand that holds the microphone. Unless you work strictly with hand-held and lavalier mikes, you'll need a stand of some kind for each microphone.

Most microphones have a standard-thread female mount designed to accept a "stand" or means of support. The bare minimum is a desk stand, a weighted metal base with a piece chromium-plated pipe threaded on each end to mate with the microphone and base (see Fig. 6-7).

The next step up is an adjustable table microphone with a velvet-action lock on it. From there we graduate to a floor stand, also with an adjustable velvet-action lock (see Fig. 6-8). The basic idea of this so-called "velvet action" is to allow you to adjust the microphone while it's being used. In practice, this isn't a very good idea since the noise of your fiddling with the mike stand will carry acoustically through the pipe and up into the mike to be recorded as an awful lot of aggravating rumbling noise.

Next in the accessory parade is the vibration or shock mount and there are several different kinds available. The simplest type screws onto the microphone stand then the mike screws onto the shock mount. This works fine for an upright microphone, but there are other cases where it won't work as well. Some microphones have built-in vibration/shock mounts.

Fig. 6-7. This desk stand by Atlas is handy for voice recording at a table or on a desktop. Mike is an Electro-Voice 631 omni.

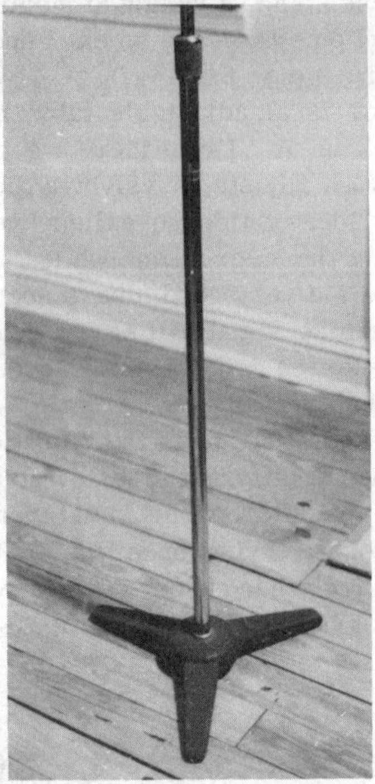

Fig. 6-8. The Atlas floor stand above has "velvet action" height adjustment. Don't ever adjust this while making a recording. It'll come over like so much thunder.

ADD A BOOM

Another item is the "baby" boom attachment, an accessory that will probably cost less than the floor stand but one that turns your setup into a more professional installation. The boom attaches to the top of an adjustable floorstand, with the microphone at one end and a counterweight at the other end (see Fig. 6-9).

Boom attachments should be adjusted with a pair of pliers —not by hand. While they have thumb screws on them, they're generally very hard to loosen, especially if you tightened them too much last time around. A word of caution: Make sure those counterweights are on good and tight, and in case you're inclined to going around in stocking feet when setting up for recording sessions, don't. (This lesson was learned from a badly bruised great toe.)

MIXING MIKES

An almost absolute must for the serious recording enthusiast is a "mixer." The mixer simply takes the outputs of two or more microphones and combines them into one channel (see Fig. 6-10). A stereo mixer does the same for two channels separately, so you can use two microphones on the left channel and two or three microphones on the right channel and adjust the relative volume for the best possible effect (see Fig. 6-11).

A simpler mixer is nothing more than a coupler, a variable resistor or potentiometer in each microphone line tied

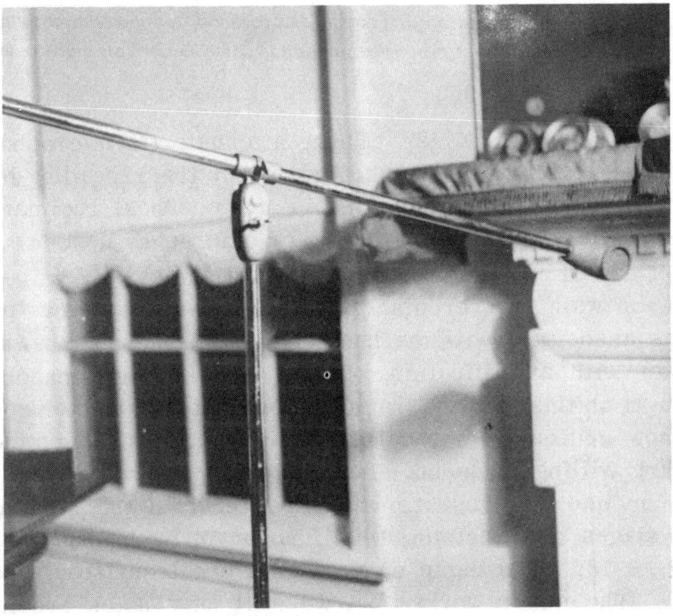

Fig. 6-9. An Atlas baby boom attachment converts any floor stand into a studio boom stand—very useful for getting that mike into difficult-to-reach places.

Fig. 6-10. Radio station mixing console accommodates many microphones, mixing them in the exact proportion needed. Similar consoles are used for professional recording.

together (see Fig. 6-12). These attenuate the microphone's signal, meaning you have to "goose" up the recording level, thus degrading your signal-to-noise ratio. These rudimentary mixers are suitable as a stopgap or emergency measure, but don't expect professional results with them.

If you want to overcome signal-to-noise and level problems associated with the basic mixer, then you must use a mixer with amplification. There are several schools of thought on this; the one that ultimately gets your vote depends on how critical your requirements are and how much money you're willing to spend. The simplest powered mixer has two or more microphone inputs per channel and has one or two stages of transistor amplification in each channel (see Fig. 6-13). The basic concept of a one-transistor type is fine. The mixer works, but we still have noise problems. The only way to get rid of the noise is to put at least one stage of amplification ahead of the voume control in each channel, then mix the signals together and put in a couple of

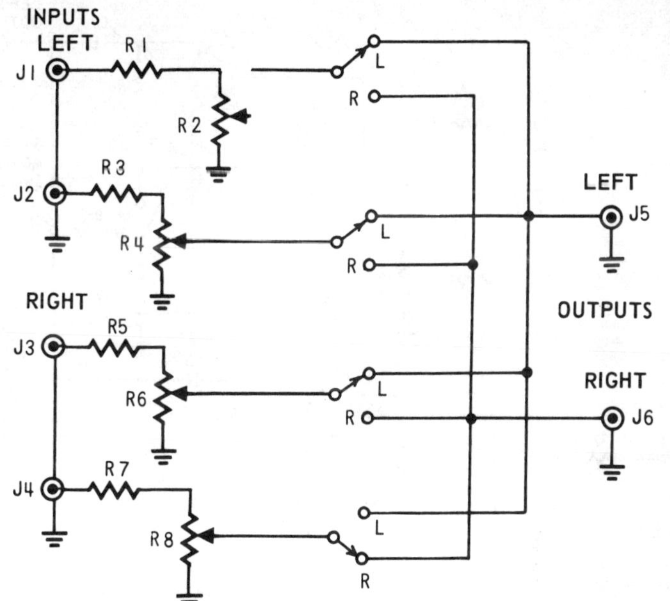

Fig. 6-11. This stereo mixer circuit with switchable inputs lets you select the left or right channel for each mike input with a SPDT switch. In setup shown, three mikes, plugged into J1, J2, and J3 all go to the left channel, while just one mike at J4 goes to the right channel.

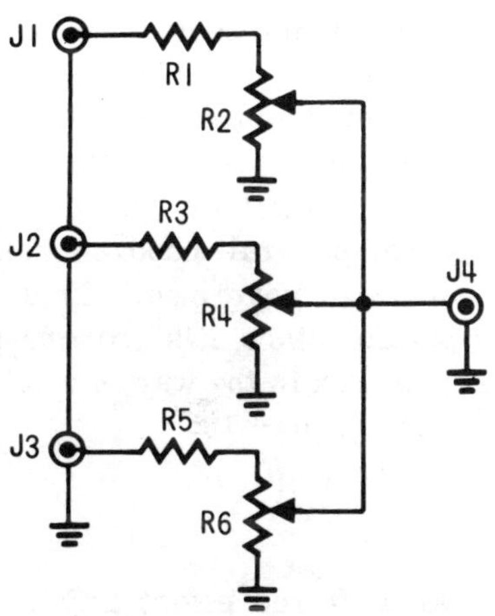

Fig. 6-12. A basic mixer is simply a series of mike inputs with a variable resistor (potentiometer or volume control) in the line, with all center wipers tied together.

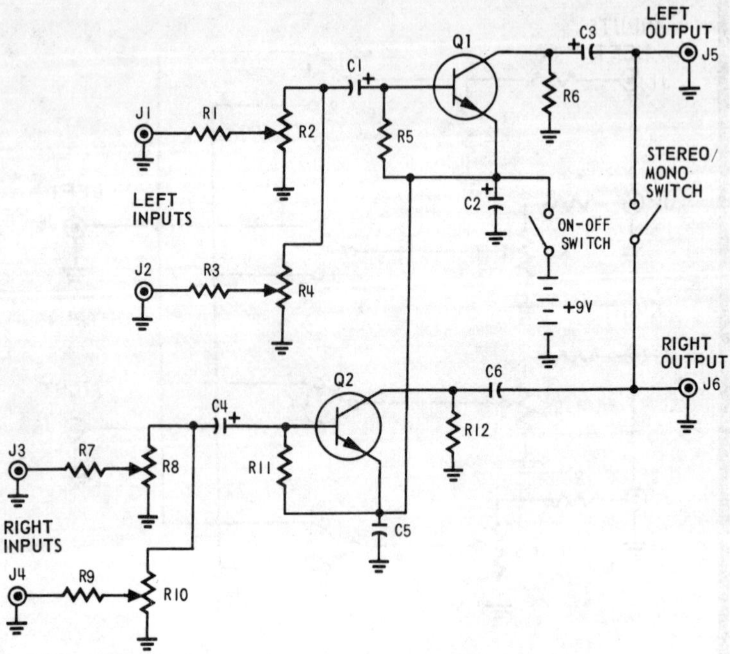

Fig. 6-13. Adding transistor amplification to a simple mixer overcomes noise and signal loss in resistor networks. This is a schematic of the Switchcraft 4-input stereo/mono mixer.

more stages of amplification. This gives a nice clean noise-free signal, but the mixer has suddenly become more expensive and more difficult to build.

A rudimentary mixer without any amplification will cost from \$5 to \$10. Transistorized units start at about \$10. You can buy monophonic mixers and pair them up for stereo if you like. You can easily spend several hundred dollars for a good facility, or you can build your own. This last course is probably the least painful since it's the cheapest and will give you exactly what you want in the way of a mixer. Some mixer circuits are shown in Chapter 10.

RECORDING IN STEREO

There are several ways to get a stereo effect in your recording. The professional audio engineer prefers to use eight or more microphones, all recorded separately (see Fig. 6-14). He'll then mix them together to get the effect that he wants. That's fine, but you don't have an 8-channel stereo tape recorder and you don't have eight microphones. However, you want good stereo separation, so use directional microphones. This means your basic microphones should be a matched pair of cardioid dynamics. Any other type of microphone will be a waste at this point.

Once you have your matched pair, you can go on and buy a single omnidirectional mike for center fill or to add onto the left or right channel with a mixer. You can go into many other microphones from there, but remember that basic pair of mikes if you want good stereo separation. Placing the microphones is another matter. You can get good stereo separation and a good stereo effect with the microphones placed only a couple of feet apart and very close to the subject, or you might have them placed 10 to 20 feet apart in

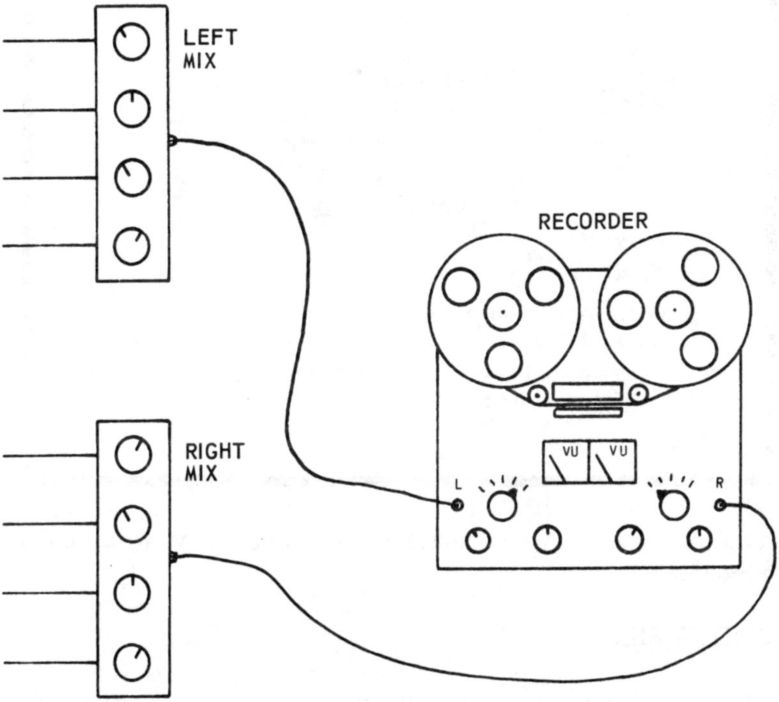

Fig. 6-14. Using two mono mixers for stereo is a common practice. This is what the basic setup looks like, schematically.

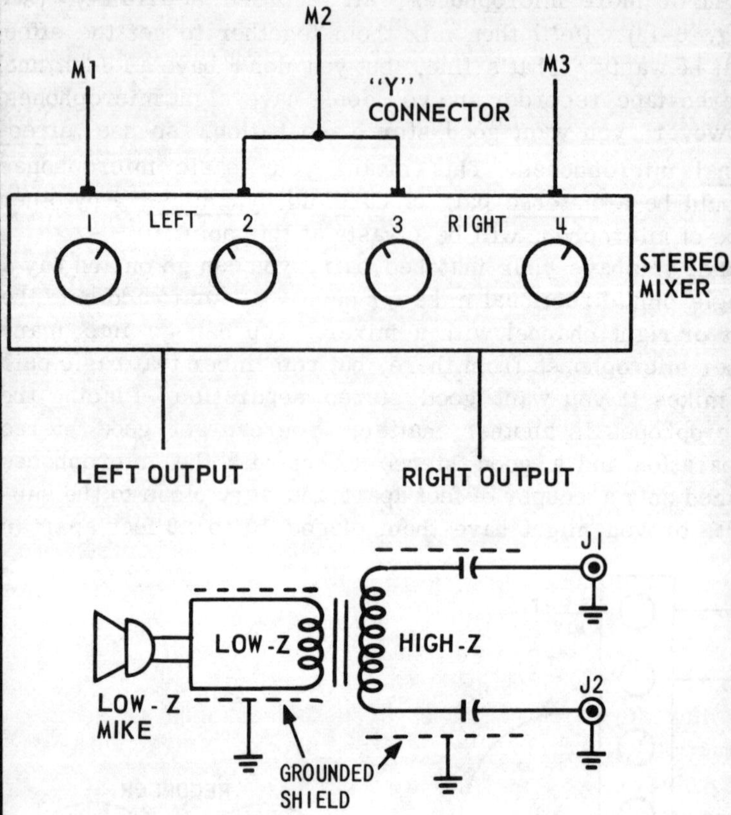

Fig. 6-15. (A) One way of adding a center-fill mike with a simple mixer. Mike M2 feeds both inputs 2 and 3 in a 4-input mixer through a "Y" connector. (B) Another method of adding a center-fill mike with no relative volume control settings. This method is very inflexible.

a large room. The ultimate result depends on what you're recording.

CENTER FILL

If you start to work with larger vocal groups and with a variety of instruments, you'll find that two microphones are no longer enough. The first thing you'll need is a mike for a center fill. This isn't a simple nut to crack. You'll have to use a mixer and set up a matching transformer to split the mike output evenly between the two channels (see Fig. 6-15).

Ideally, a center mike should be set up with a "panpot" control, the kind used in professional recording studios. With a "panpot" you can locate the apparent position of the center microphone anywhere from left to right or any point in between. A simple panpot is easy and inexpensive to build and will help in a lot of your recording sessions, especially when that third wheel is there to be added. For more on the panpot, see Chapter 17.

ROOM ACOUSTICS

A perennial problem is the acoustic one. What do you do in a room that's not designed as a recording studio but is being used for one? The most usual place for home recording is the living room with its heavy, overstuffed furniture, thick drapes, and (hopefully) heavy, padded carpeting. This is fine—the room is nice and dead acoustically, just about the same as any recording studio worth its salt. But a recording studio can have its acoustics changed to meet the situation with various devices and props. Not so with a living room.

One thing you can do in a living room is open the drapes to get some reverberation from the window or from the wall behind them, and this in itself is a very useful trick. Another thing you can do in a living room is to try opening or closing doors and windows to see what effect this has on the reverberation time. This will take some experimenting.

If you want to add some reverberation to a room that's accoustically dead, there's only one way to do—cheat. "Cheating" involves the same method used by professional recording engineers. They use an electro-mechanical device known as a reverb system, a device that adds reverberation to the sound artifically (see Fig. 6-16). It's a beautiful gimmick because you can adjust the amount of reverberation that you want. You can add just ever so little or you can add so much echo that it sounds as though performers are in a huge, cavernous hall.

You need two reverb units for stereo (one for each channel)

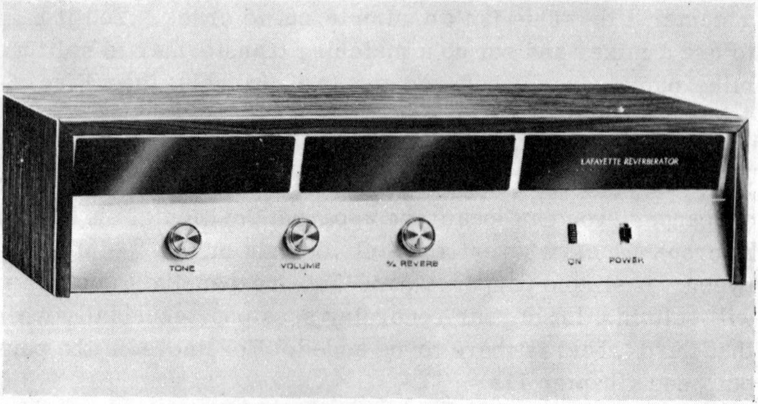

Fig. 6-16. A reverb unit, such as pictured here, will add an extra dimension to your recordings.

or you can get by with one by just putting the reverb on one side. Trouble is, this may make the other channel sound dead. By using two units you can balance the reverb and get a very full, rich-sounding stereo effect. You can do this after you've already made your master recording by adding reverberation in the duplicate, trying out different degrees, until you get the effect that you want. The difficulty is that you may have several instruments on the same channel and you may want to add the reverb only to one instrument or one vocalist. This is a judgment you must make at the recording session. If you feel you can add the reverb later on, fine. If not, add it while you're making your recording in the particular microphone line that needs it.

If you're at all serious about your recording sessions, then you should provide yourself with a serious tape hobbyist's recorder. This type of recorder has features normally not found on the lower-priced units. One very important feature is a third head. You'll recall that back in Chapter 1, we talked about the different kinds of heads. We had recording heads which were usually the same heads simply switched electrically for the different functions, and we also had an erase head.

The basic machine works fine with just two heads—one a dual-function job. But the professional machine has three heads: the erase head, the recording head, and a separate playback head—each one optimized to work best at its intended function. There's another big plus to having three heads. You can listen to the tape, preferably on a good pair of stereo headphones, while it's being recorded. The tape passes over the erase head, then over the recording head, then over the playback head in that order. So, a split second after the tape has been recorded, you can listen to the playback and see how it's taking. This way you can make any adjustments that you need on the spot. If you find that the take is no good, you have an opportunity to do it over without subjecting everyone to a long playback and rerun through the tape.

Monitoring this way raises an interesting question which is largely a philosophical one among the uninitiated. The question is, how important is it to "ride the gain"? Riding the gain means sitting there watching those needles with your hands on the recording level controls and adjusting them if the level gets excessively high or excessively low. Riding the gain in a professional recording session is an absolute must. If it's just a casual get-together, and you're not too concerned about the quality of the tape, then forget about riding the gain. But once the bug of perfectionism has bitten you, ride it you will.

And there you'll sit with those big stereo headphones clamped on your head looking ever so much like a Martian, looking completely oblivious to what's going on while the recording session is in progress. Fact of the matter is you'll probably be having more fun than anybody except the musicians, since you'll be an active participant in making the recording, and the quality of that recording hinges on exactly what you do every second of that session. Be prepared to take some ribbing, though. Most people don't understand tape nuts.

SOME DOS AND DON'TS

Since a lot of the basics in microphone handling and placement is necessarily a trial-and-error operation, here are a couple of dos and don'ts you might bear in mind: Don't place the microphone 10 feet away from a singer and expect to get a good recording. Sure, you can crank up the volume control, but if you listen on the stereo headphones and monitor the tape you'll find that you've got a very hollow "echoey" sound. This is not professional recording; this is amateurism at its worst. If you are going to record this way, don't bother to get better microphones.

You may also want to try a few tricks for getting candid recordings. This is especially important for the small fry, since some of them may not act natural if there's a microphone being held under their noses. To do this, you'll need an accomplice or someone to help the kids over to a hidden microphone. Many mikes are very small and can be disguised as dolls or little figurines that the children can be encouraged to talk to. Sometimes, this won't be necessary, because some kids are just natural hams who are delighted to grab the microphone and ham it up for you, but all that noise from the mike handling! And if you monitor it with headphones, goodbye eardrums!

There are many special techniques that you can work out and many that will suggest themselves as time goes on. It takes a certain amount of practice, plus a certain amount of skill, patience, and some plain dumb luck.

Chapter 7

Wireless Microphones

The wireless microphone can be a boon or it can be a disaster, depending on just what you need as a tape recorder accessory. Basically, the wireless mike is nothing more than a miniature radio transmitter—not just a microphone without a wire as the name might imply. Since you have a transmitter, you must also have a receiver. Usually, any of the "garden variety" FM radios will do.

The wireless mike comes in many shapes, sizes and forms (see Fig. 7-1). Almost all of them have one thing in common—they broadcast in the standard FM band. To get the most use out of the microphone, you'll have to retune it to a vacant spot on the FM dial.

In its cheapest form, these units come already equipped with a crystal microphone element, completely eliminating your choice of microphone type. At the other end of the scale are the rather expensive jobs that use a good quality dynamic along with a no-expense-spared transmitter. These can run as high as $300 or $400. Carried to the ultimate extreme, there are wireless mikes used in the broadcasting industry which can run into thousands of dollars (see Fig. 7-2). These wireless microphones do not broadcast on the FM band. They're made for use by pros and they work at microwave radio frequencies—around 2.5 GigaHertz (2500 MegaHertz). This kind of operation demands very expensive equipment.

BUILDING A WIRELESS MIKE

It's quite possible to build a wireless attachment for your favorite microphone at a cost of about $5.00 or less. The

Fig. 7-1. A wireless microphone frees the performer to move about without regard to cables trailing on the floor.

easiest way to do this is to use an encapsulated module designed for this purpose, such as the Carl Cordover type FM-1 or FM-2 (see Fig. 7-3). The two different modules are both FM transmitters, one made for high-impedance mikes, the other for low-impedance jobs.

The module's manufacturer indicates that the frequency can be changed by placing a coin near the transmitter or altering the length of the antenna lead. Either method is very cumbersome and difficult to control. A much better way is to install a 25- to 50-picofarad adjustable capacitor between the tank circuit and the battery supply (see Fig. 7-4).

Also at the low end of the price scale are wireless mike kits available from EICO (see Fig. 7-5). Even working at their best, the range is limited as on most low-cost mikes. These mikes use a built-in crystal element, so the frequency bandwidth is very limited.

The circuitry used in most wireless microphones is rela-

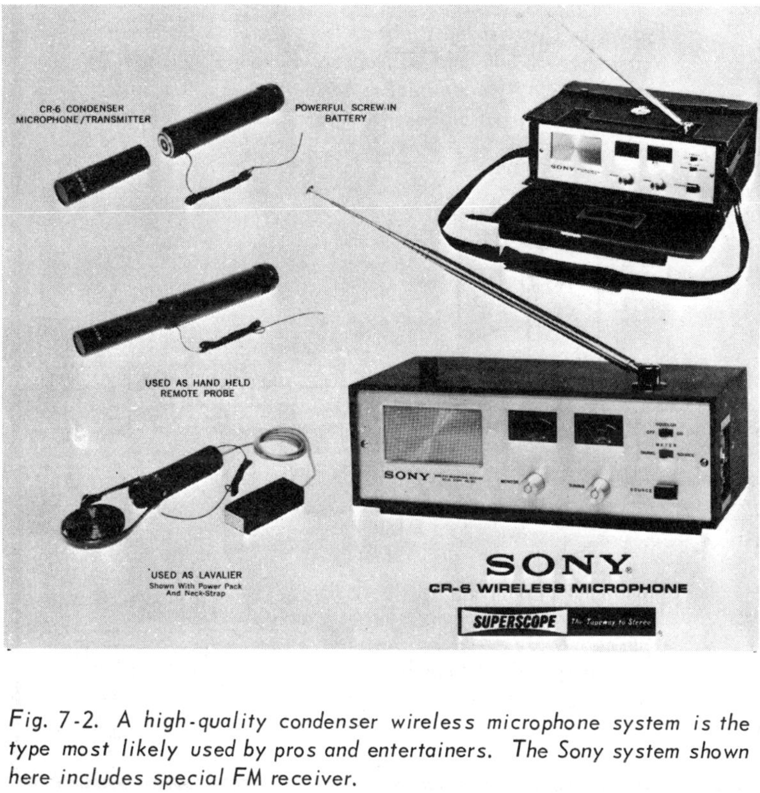

Fig. 7-2. A high-quality condenser wireless microphone system is the type most likely used by pros and entertainers. The Sony system shown here includes special FM receiver.

Fig. 7-3. Home-made wireless FM mike uses Cordover transmitter module and will accept a variety of microphone types. A single penlite battery powers this unit, and total construction cost is about $5.

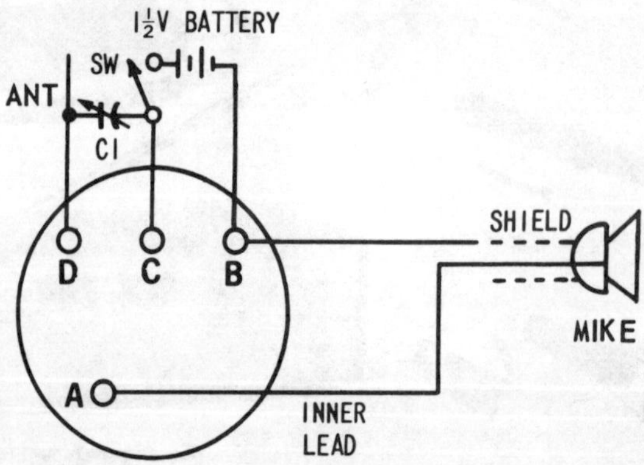

Fig. 7-4. Basic hookup for an FM wireless microphone module. Add a 25-50 pfd trimmer capacitor (C1) between the "antenna" lead and positive battery terminal for tuning.

tively simple and straightforward, generally using only three or four transistors. This is a low-power transmission, and it has to be; otherwise there'd be too much interference with neighboring FM receivers. Generally, you can expect a useful range of 100 to 200 feet with most wireless mikes. Under ideal conditions, you can get more distance but not necessarily with good fidelity.

FREQUENCY RESPONSE

The frequency bandwidth that you can get from an FM mike depends on each component in the chain: the microphone's frequency response, the undistorted bandwidth of the FM transmitter; the discrimination characteristics of the receiver; the type of hookup from the receiver to the tape recorder (see Fig. 7-6). If you're working with a battery-operated portable recorder and a little battery transistor FM radio, then naturally you're not going to get as good fidelity as you would with a more sophisticated rig.

In the living room, you can use the FM tuner in your stereo

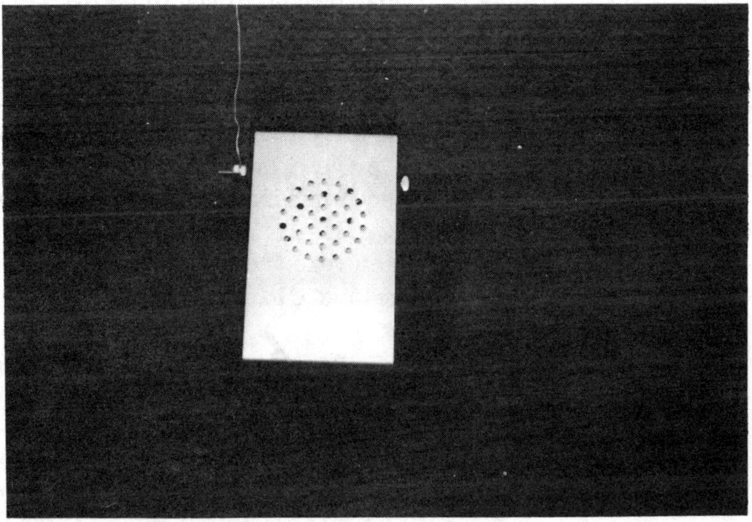

Fig. 7-5. Wireless FM mike in kit form by EICO sells for $10. It has a crystal microphone element built in, but an external mike could be used by modifying the unit.

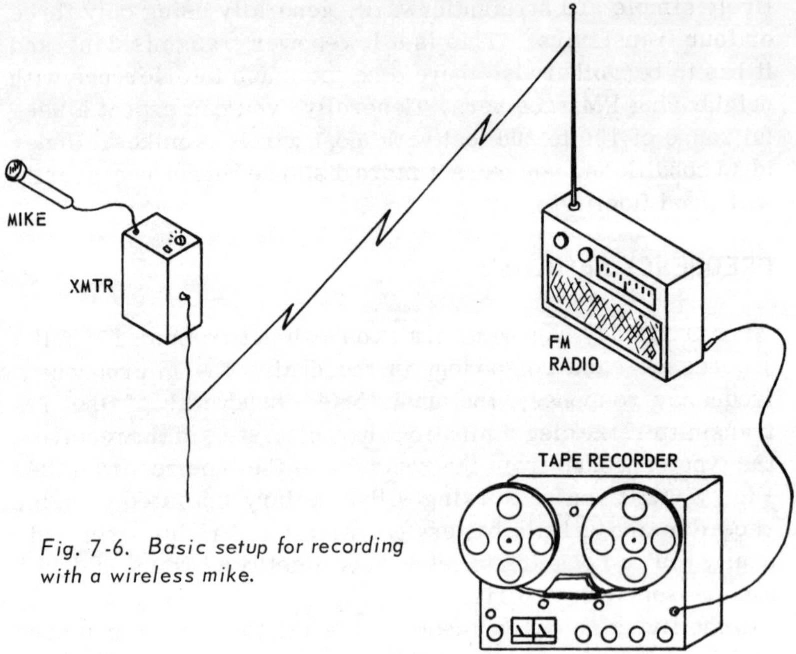

Fig. 7-6. Basic setup for recording with a wireless mike.

rig piped right into your stereo tape deck. Here, you're optimizing the receiving end, and it's up to you to do the same for the transmitting end. Surprisingly, some of the best low-cost FM mikes come with certain model Japanese tape recorders that have built-in FM radios. But it's difficult to purchase one of these microphones separately and they're certainly hard to duplicate at a comparable price. The most popular mike of this type is a companion piece for certain model Crown tape recorders.

Other wireless mike manufacturers include Kinematix, with a self-contained crystal mike and Channel Master (see Fig. 7-7).

The legality of using wireless FM microphones is something that has never been fully settled. The FCC doesn't like them too much but they do fall within a provision of the Communications Act that permits very low-powered FM transmitters in the standard FM broadcast band. In order to be acceptable and legal, your microphone should be type-approved. This means that the manufacturer must submit to the FCC for testing and approval a working sample of a production model that he intends to sell. This is obviously impossible for home-built microphones and individual kits. Here, you're treading on the edge of the law and should try to check for possible type approvals before going ahead with anything.

THE VERSATILE FM MIKE

Any wireless microphone is potentially a wireless "bug." A "bug" is simply an FM wireless mike that has been hidden. This is the only real distinction between surveillance devices and ordinary wireless mikes. You'll read more about this type of device in Chapter 21.

One other thing to bear in mind: As with transistor radios, the quality and transmission range of your microphone depends on the condition of the batteries. Keep them in topnotch shape. Also, be careful of antenna orientation when you're using the mike. In some models, antenna orientation—both at the mike and the receiver—can make a lot of difference.

Where to use the FM mike? The actual application depends on what kind of recording you're going to do. In one very difficult recording session in a church, we found ourselves faced with the dilemma of trying to tape all of the priest's intonations while he moved from one part of the altar to another, never

Fig. 7-7. One of the most effective low-cost mikes, this two-penlite powered model is available only with certain Crown tape recorders with a built-in FM radio.

staying in one place very long. This was a place that called for an FM microphone; no other technique would do the job.

In other situations, performers very often benefit from the cordless operating ease of the wireless microphone. The same is true of participants who must be extremely mobile and frequently move beyond the reaches of microphone cables.

A very effective type of wireless setup is to use a lavalier mike plugged into a "body" transmitter in the user's pocket. This way the microphone is always oriented properly for best pickup and the user can use it in hands-off operation.

Next time you have a difficult recording session ahead of you, try using a wireless mike. It can add a lot of versatility to your total microphone setup and can be a lot of fun, especially at parties.

Chapter 8

Amplifiers & Preamps

In the course of making tape recordings, you'll eventually run into problems that you hadn't counted on. Perhaps a signal level is too low, or the tape deck that you're using on location doesn't have enough output to drive a pair of monitoring headphones or loudspeakers. And there are other specialized areas such as accessory items that you may need for your tape recorder. All of these difficulties have one thing in common—they require some kind of amplification.

Probably the most common problem today with the new generation of transistor tape recorders is the fact that the input amplifier doesn't have enough gain for all the types of program sources you may want to use with it; for example, when you have to crank up the recording level control all the way to get any indication of usable volume on the VU meter.

This problem is especially apparent when you're recording from a low-level phonograph pickup, even though it first goes through the preamp stages of your stereo amplifier. What to do? The answer is relatively simple: install another preamp in the tape recorder line. The preamp can be set up in one of several ways. The most obvious one is to build it into a Minibox and plug the tape output jumpers into the preamp input and then plug the preamp output into the tape recorder line input. A complete stereo circuit fits nicely into a simple Minibox (see Fig. 8-1).

INEXPENSIVE PREAMP

Heart of the stereo preamp is transistor type 2N3391A. This is a low-cost (45¢) quality audio transistor made by General

Fig. 8-1. This simple stereo preamp circuit uses a 2N3391A transistor in each channel for general purpose use.

Electric. It's rather tiny and can be difficult to handle, but it works beautifully in the circuit and should not be any trouble at all to hook up on a pre-drilled phenolic circuit board. The base diagram is shown in Fig. 8-2.

The input for the left channel is applied to jack J1 (see Fig. 8-2). The signal goes through C1, a 10-mfd electrolytic, to the base of Q1, the 2N3391A transistor. Bias is provided by R1 and R2. The emitter is grounded through 680-ohm resistor R3, bypassed by C2—a 25-mfd miniature electrolytic. The collector gets its B-plus through R4, a 20K resistor, and feeds the signal output through C3, a 10-mfd capacitor, to jack J2. The right channel of the circuit works exactly the same way. The only common point is that both circuits are on the same board and get their positive voltage from the same switch and battery.

Notice that the circuit in Fig. 8-2 uses a negative ground. This is true because the transistors are types which often are easier to use than the more traditional transistors and the positive ground that they require. Using a negative ground makes the equipment compatible with commercial tape recorders and older tube-type equipment.

One way to make this preamplifier more versatile is to use a multiple switching arrangement as shown in Fig. 8-3. This switch can be a multi-gang slide switch with S1 as one of the gang. In Fig. 8-3, the power switch is S1A.

The switching arrangement shown in Fig. 8-3 requires essentially a 5-pole double-throw switch, generally not available in a slide switch, but a 2-position rotary would do the job. The triangular outlines in Fig. 8-3 represent the preamp circuit shown in Fig. 8-2. This switching arrangement is optional and you can add it to the circuit for extra convenience. When you use this switch mounted in the same Minibox with the preamp, you'll never have to unplug and replug. Just by flipping the switch you either turn the preamp on and put it in the circuit or turn the preamp off and take it out of the circuit completely. In the off position, the battery is disconnected and both preamplifiers are completely bypassed by a direct connection between J1 and J2 for the left channel and between J3 and J4 in the right channel.

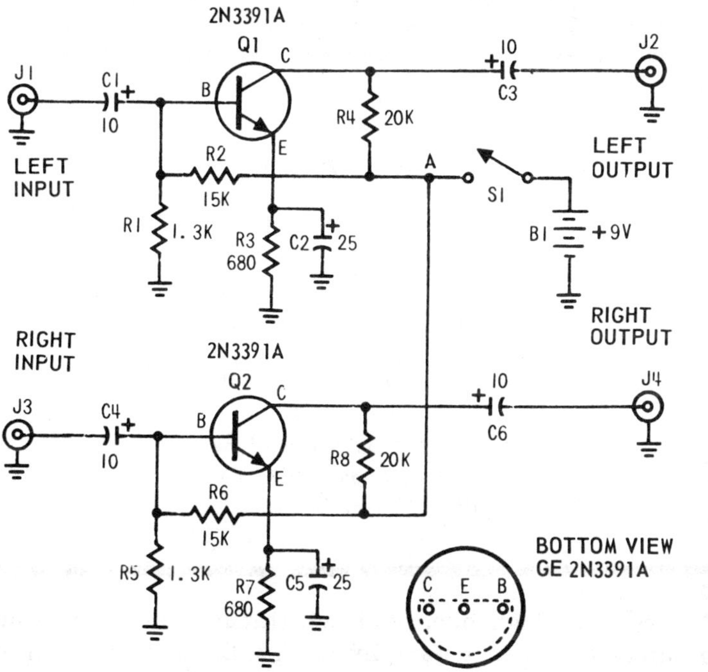

Fig. 8-2. Schematic of the simple stereo preamp shown in Fig. 8-1.

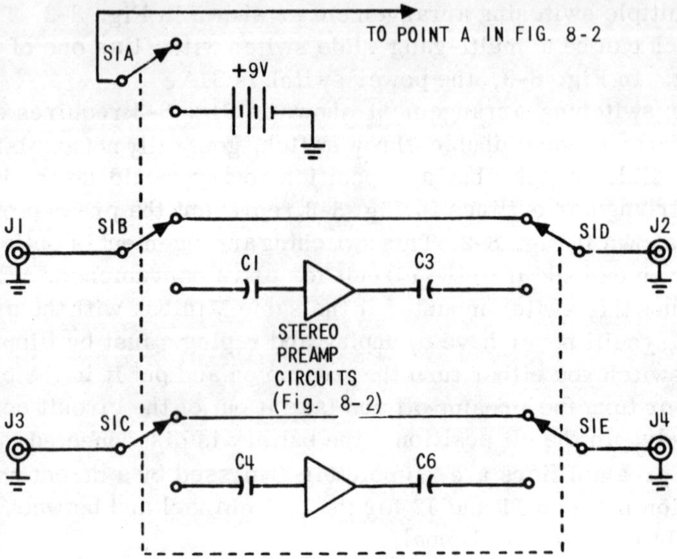

Fig. 8-3. Bypass switch arrangement for permanent preamp installation. Switch S1, a 3-pole double-throw type, turns on the power and switches both preamp channels into the circuit when needed.

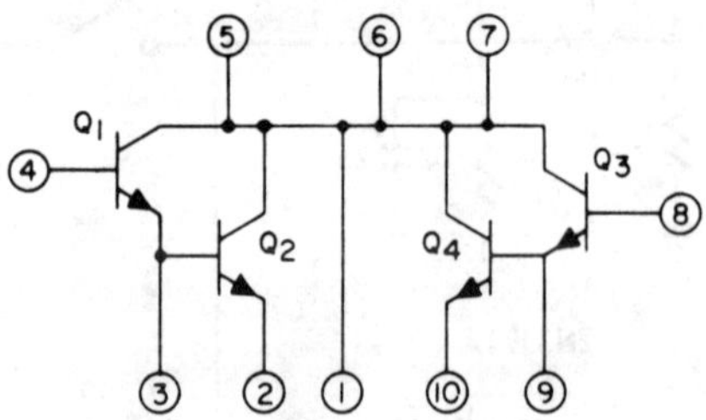

Fig. 8-4. Schematic of the dual-Darlington RCA amplifier IC type CA3036.

A very useful high-fidelity circuit that uses an IC (integrated circuit) is the Darlington amplifier. The Darlington is a direct-coupled, 2-stage solid-state audio amplifier with very low noise characteristics (see Fig. 8-4). The transistors are NPN types and the one shown here is a dual Darlington—a stereo configuration—in an integrated circuit (RCA type CA3036). This same dual-Darlington is used in the slave-driver switchbox described in Chapter 11. It's not necessary to use the Darlington in the switchbox. It can be used all on its own and has a remarkable ability to "clean up" a stereo signal on its way to the tape recorder. It also is an ideal buffer amplifier and helps match impedances and smooth out impedance mismatches.

Hooked up as a basic buffer amplifier circuit, the integrated circuit type CA3036 appears in Fig. 8-5. It is set up in a standard Minibox with jacks J1 and J3 for the left and right stereo inputs while J2 and J4 are the stereo outputs. This IC is a 10-pin type and only every other (the even-numbered) pin is used when connecting it. One way to get rid of the unused pins (the odd-numbered ones) is simply to snip them off. Snip off pins 1, 3, 5, 7, and 9 close to the case but without putting any strain on the leads. This will leave pins 2, 4, 6, 8, and 10 to be wired up as shown in Fig. 8-5.

The left input goes through C1, a 10-mfd capacitor to pin 4 of the IC. The right input from J3 similarly goes through C3 to pin 8. Biasing for the base input circuit is developed across resistors R1 and R3 tied from pins 4 and 8 to ground. R1 and R3 are each 540,000 ohms. The output is taken from pins 2

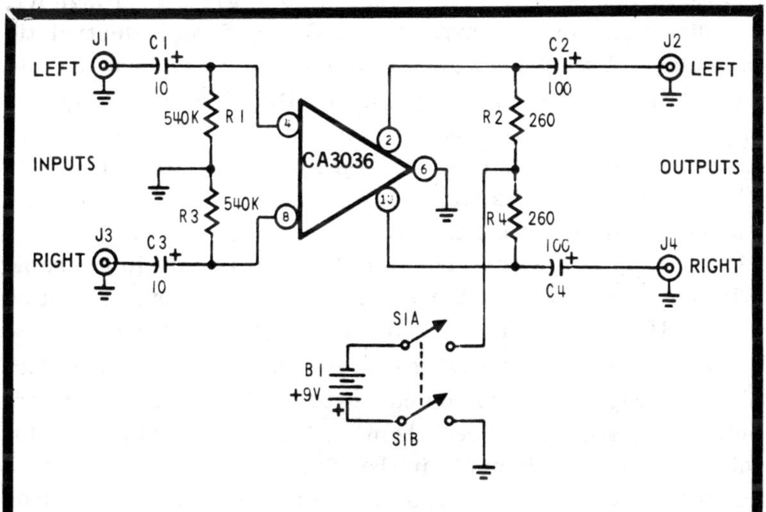

Fig. 8-5. Stereo preamp circuit using the CA3036 IC.

and 10 for the left and right channels. Both of these pins go to the B-minus supply through 260-ohm isolating resistors R2 and R4. Also taken from pins 2 and 10 are the output signals through coupling capacitors C2 and C4, each 100 mfd. Pin 6 is grounded.

A standard 9-volt transistor radio battery does the job, and both negative and positive should be switched to prevent any leakage. For this reason S1 which can be a simple slide switch, but it must be a double-pole type (DPST or DPDT).

INTEGRATED CIRCUIT PREAMP

Another preamplifier IC that you might want to try for various tape recording applications is the RCA type CA3007. If you want to set up for stereo, you'll need two, of course. The internal diagram of this IC is shown in Fig. 8-6. One of the advantages of this IC is that it provides much more gain than a single-stage preamplifier and gives you the option of adding a volume control at the output. It operates from a standard 9-volt battery in Fig. 8-7.

The two stereo channels in this preamplifier are identical. The signal input for the left channel is fed through J1 and capacitor C1, a 5-mfd electrolytic, which goes to pin 1 of the integrated circuit. Pin 3 is grounded. Pins 2, 6, 7, 11, and 12 are all connected together and all go to ground through R1, a 5,000-ohm resistor bypassed by C2, a 5-mfd electrolytic capacitor. The B-plus goes to pin 9, which is connected to the positive side of B1 through switch S1A. The output appears at pin 8, which goes to the positive side of C3, a 5-mfd electrolytic. The signal from C3 goes through R2 to ground. R2 (the volume control) is a 1,000-ohm audio-taper potentiometer. The center wiper of R2 goes to J2, the output jack.

It's a simple matter to put two of these circuits in the same Minibox with two volume controls and double input and output jacks. If you use this permanently plugged into your stereo system, it might be a good idea to use the kind of on-off switch shown in Fig. 8-3, so you can bypass the preamplifier circuits when you don't need them. There's a good reason for this: The fewer elements in the signal path when you make a recording, the better off you are, since more elements mean more of a chance for noise, distortion, and other problems to creep into the recording process. If you need the preamp, then all you'll have to do is flip the switch and it's on; when you don't need it, leave it off and it's bypassed right through the Minibox.

POWER AMPLIFIERS

There will be times when you'll need full-scale power amplification with perhaps 1/2 watt or more of power for the stereo channel. This is especially true in setting up your tape recorder for stereo headphone monitoring and for driving a pair of small monitoring loudspeakers. Generally, you don't need a very expensive or powerful amplifier for this if you just want to monitor. For that matter, you can even drive a fairly small high-fidelity loudspeaker system using some of the amplifiers available today. A speaker such as the Electro-

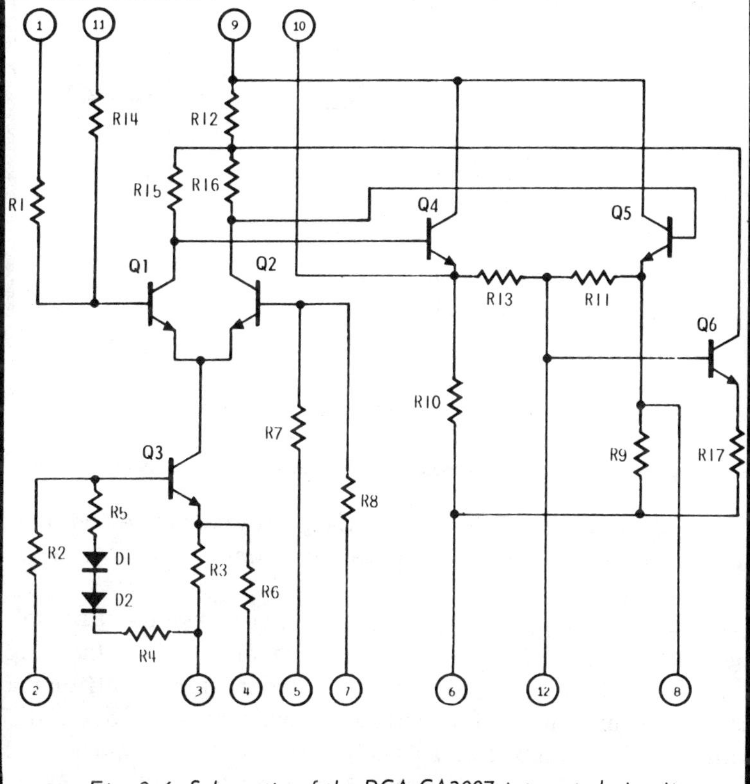

Fig. 8-6. Schematic of the RCA CA3007 integrated circuit.

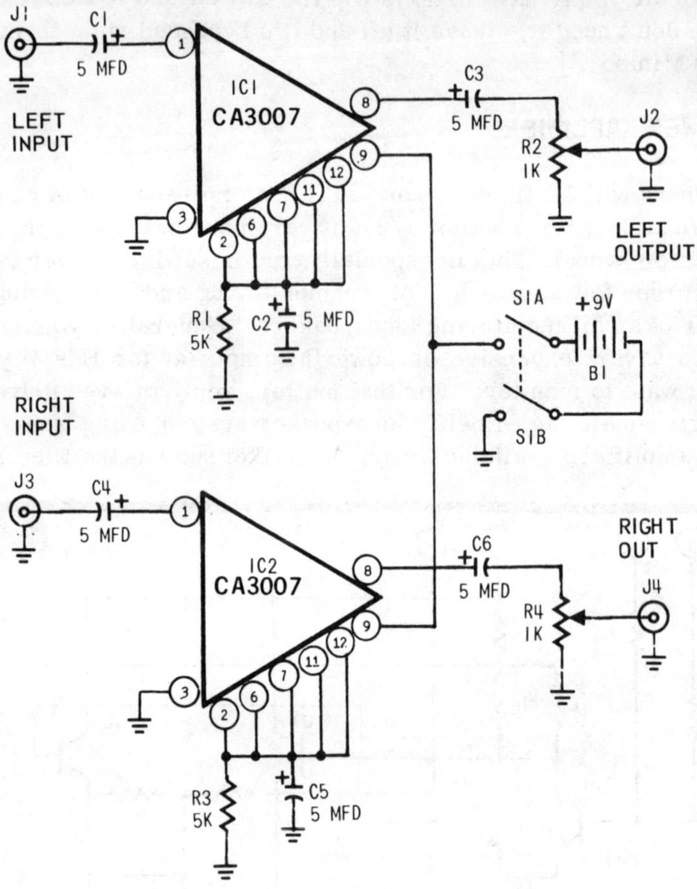

Fig. 8-7. Two CA3007s wired as a stereo preamp with variable output.

Voice EV-Eight has an excellent frequency response and sound quality and yet will operate quite well from only a watt or two of audio power.

A relatively simple way to get this power is to use prepackaged power amplifiers such as the Lafayette PK-544 (see Fig. 8-8). This is a push-pull output all-transistor amplifier that has an 8-ohm output impedance. Power output is 360 milliwatts, and the amplifier sells for $5.85. A 1-watt push-pull amplifier from Lafayette costs $6.95 and a 3-watt job is $7.95. If you want full stereo capability at minimal cost, also in the

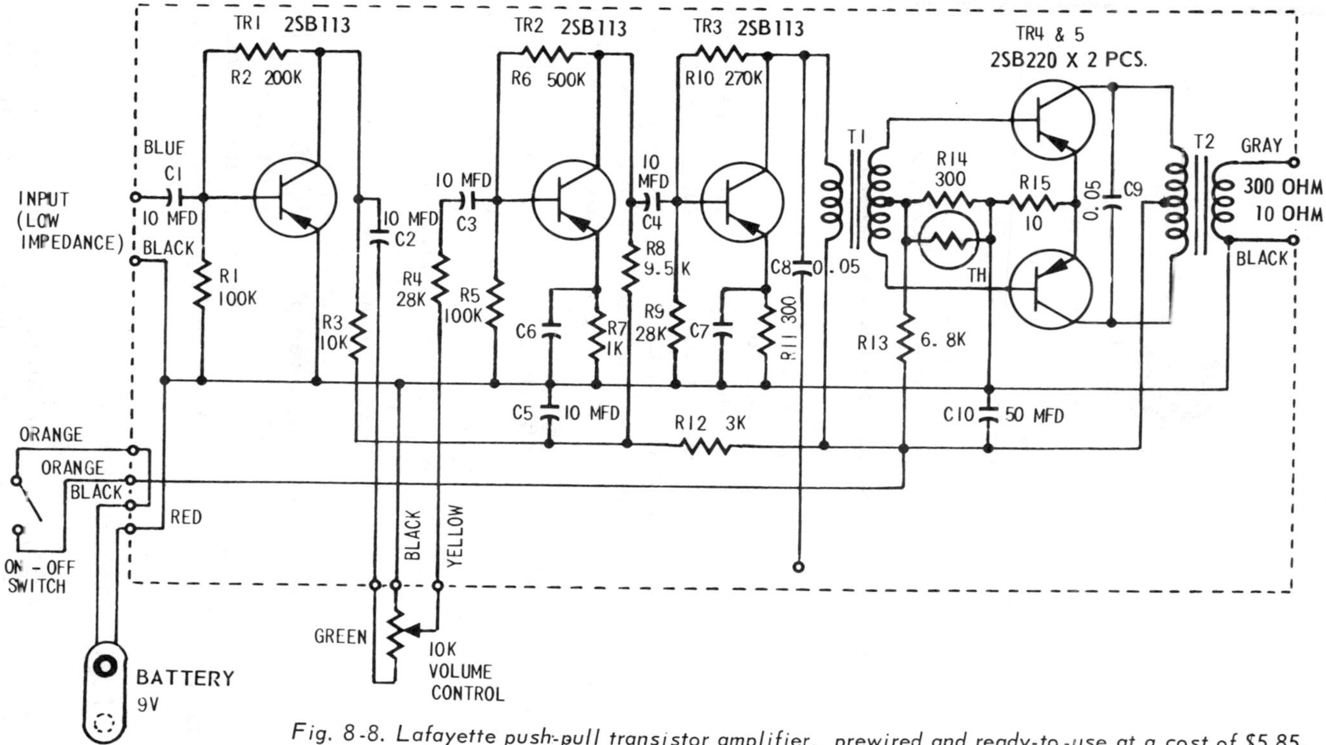

Fig. 8-8. Lafayette push-pull transistor amplifier, prewired and ready-to-use at a cost of $5.85.

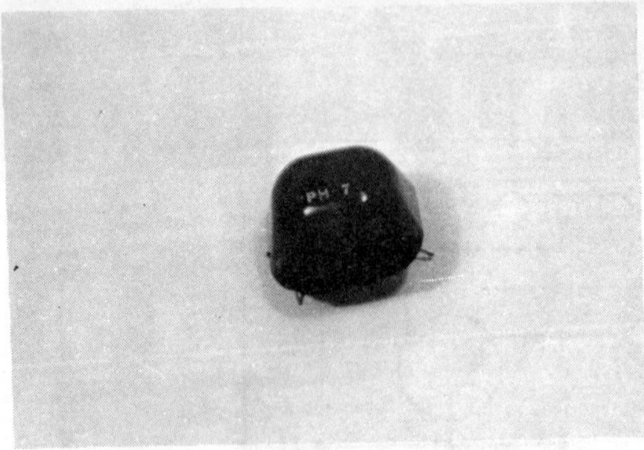

Fig. 8-9. This Cordover encapsulated amplifier module sells for $3 to $3.50; all you have to do is connect four wires.

Lafayette catalog is a 5-watt stereo amplifier—2 1/2 watts per channel—and this one is tagged at $10.95.

Most of these amplifiers operate from a standard 9-volt battery, although some may require higher voltages—especially the 5-watt amplifier; this one takes a 15-volt DC supply. In some cases it may be advisable to build a separate power supply for these amplifiers to save battery costs. This is especially desirable if they're going to be used where AC is readily available.

Another way to amplify is with a sealed, encapsulated module. A Cordover module for phone amplification (which will work well here) sells for $3 to $3.50 and will drive an 8-ohm loudspeaker at about one watt (Fig. 8-9). This module takes a 6-volt supply, but the manufacturer recommends a lantern battery because of high current consumption. With modules like this, an AC power supply is just about a must.

Cordover also makes a stereo phono amplifier module, Model STX, and this one sells for $5. It's the same as the phono module except that there are two of them in one package, and again an AC power supply would probably save a lot of batteries.

IC POWER AMPLIFIERS

If you want to do a little additional work and use some of the latest sophisticated circuitry, you can very easily put together an integrated circuit stereo power amplifier. The integrated circuit in this case is an RCA type CA3020A. This IC has an output of 1/4 watt and is a dandy (see Fig. 8-10).

Unfortunately, like most ICs, it requires a certain amount of external hardware. Parts needed for this amplifier are shown in Fig. 8-11. You can build a monophonic version simply by using 1/2 of the schematic in Fig. 8-11 or use two ICs in the setup shown. Looking at just the left channel, the signal input is applied through jack J1 and capacitor C1, a 1-mfd electrolytic connected to the input transistor base at pin 10. R1 is a 5,000-ohm audio taper potentiometer which is the volume control. Precision resistor R3 is 0.62 ohms and provides just the right amount of isolation while grounding pins 5 and 6.

The push-pull output is present at pins 4 and 7 which go to a push-pull type output transformer, such as Lafayette's Argonne type 174 or equivalent. The type 174 has a 4-ohm secondary. Other transformers may be substituted to get a higher impedance for the loudspeaker. A finned heat radiator force-fitted on the IC case is a must.

A double-pole double-throw switch (S1) is used to disconnect

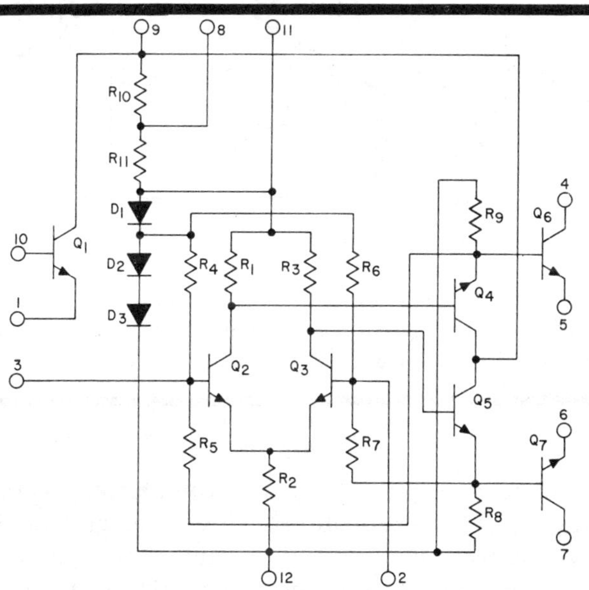

Fig. 8-10. Internal wiring of the CA3020A integrated circuit amplifier.

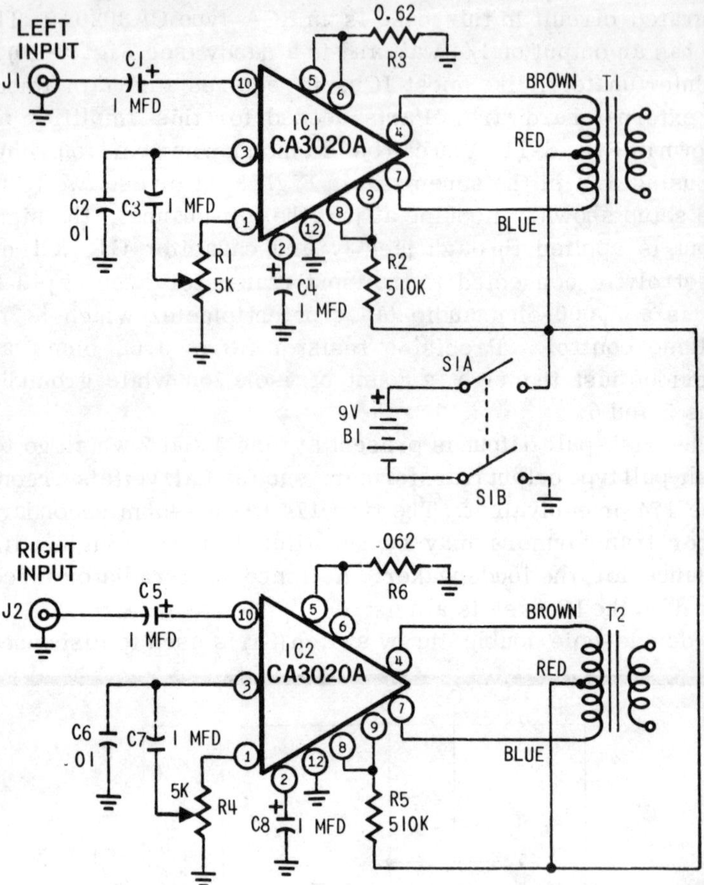

Fig. 8-11. The CA3020A intgrated circuit wired into a stereo amplifier circuit. These ICs will provide adequate power to drive small loudspeakers.

the entire battery supply which is a 9-volt standard transistor radio battery. This particular integrated circuit is the 12-pin type. Use the usual amount of caution that you would in soldering transistor leads. Remember, this is a little more expensive than a single transistor, but it can get ruined just as fast.

Chapter 9

Microphone Mixers

There certainly will come a time when you will want to use more than one microphone per stereo channel. It may be because you have instrumentalists who need a certain kind of attenuation or mixed-in balance for their instruments; it may be because you have too many people for a 2-microphone setup, or for any number of reasons. No matter what the reason, if you want to use more than one mike per channel, you'll have to use a mixer of some kind.

What the mixer does is to take the signal from two microphones or more, mix them together and "blend" them into a single channel. The simplest mixers adjust the relative loudness of the two microphones with a variable resistor (potentiometer) in the line (see Fig. 9-1). While this elementary system works pretty well, it naturally attenuates the sound on one or both channels, and you can have a noise problem as a side effect. To make up for this attenuation, it's usually a good idea to include some kind of amplifier circuit in the mixer. This brings the sound level back to a usable signal voltage and compensates for the attenuation losses in the mixer controls.

AMPLIFIER MIXERS

There are several schools of thought on just how to do this. Most inexpensive mixers simply use one or two stages of amplification after the signals have been combined. This is an easy and an effective way of doing it (see Fig. 9-2). But professional audio engineers will tell you, and rightfully so, that the noise level has appreciably worsened in the process. The reason for this, they will say, is that each input channel should have some amplification ahead of the attenuator or volume

Fig. 9-1. The simplest kind of mixer, a two-input mono resistance bridge snugs into palm of the hand above. Inside that metal case are two potentiometer volume controls for the two inputs—R2 and R4, with their center wipers tied together to form a single output.

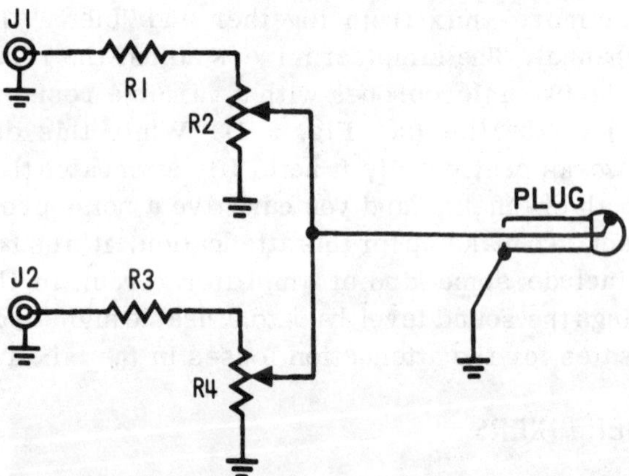

control. This is a very valid point of view, and just this type of circuit arrangement is used in the professional audio mixer. The trouble is that adding this amplification in each input channel can raise the cost of the equipment—especially if expensive FET (field-effect transistor) inputs are used.

One of the decisions you'll have to make about a mixer is the

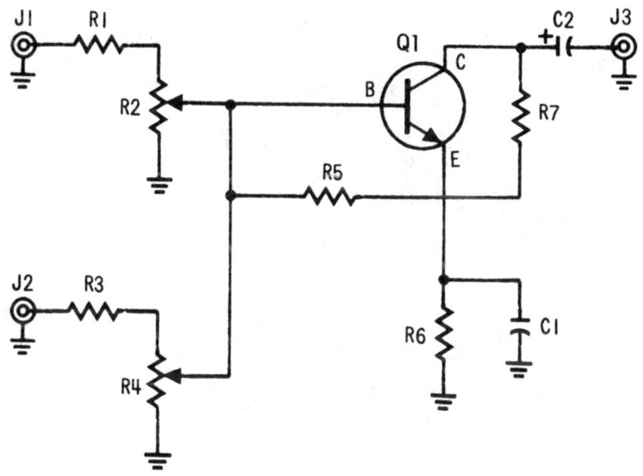

Fig. 9-2. To the simple mixer circuit in Fig. 9-1 a single transistor amplifier stage has been added here to overcome attenuation and noise.

Fig. 9-3. Two inputs per stereo channel or four mono inputs are provided by this low-cost Switchcraft mixer, which has a single stage of amplification in each stereo channel.

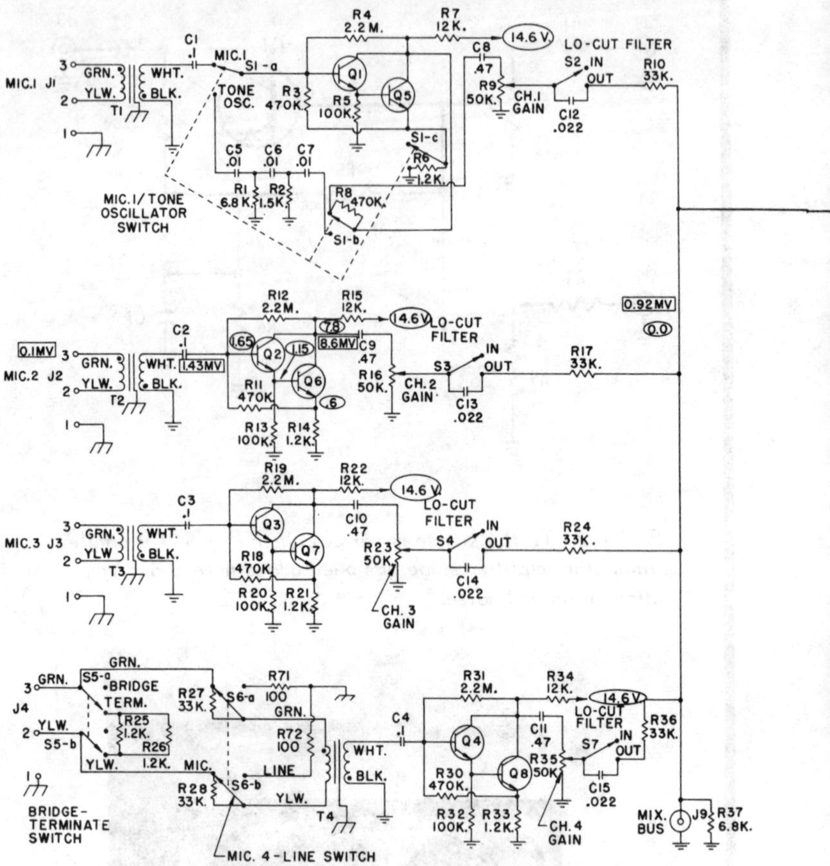
MIC.1 J1
GRN.
YLW.
WHT.
BLK.
T1
C1 .1
MIC.1
S1-a
TONE OSC.
R4 2.2M.
R7 12K.
14.6 V.
R3 470K.
Q1
R5 100K.
Q5
C8 .47
R9 50K.
CH.1 GAIN
LO-CUT FILTER
S2 IN OUT
R10 33K.
C12 .022
S1-c
R6 1.2K.
C5 .01
C6 .01
C7 .01
R1 6.8K.
R2 1.5K.
R8 470K.
S1-b
MIC.1/TONE OSCILLATOR SWITCH
0.92MV
0.0
0.1MV
MIC.2 J2
T2
C2 .1
1.43MV
R12 2.2M.
R15 12K.
14.6V
LO-CUT FILTER
1.65
Q2
1.15
7.8
8.6MV
C9 .47
R11 470K.
Q6
.6
R13 100K.
R14 1.2K.
R16 50K.
CH.2 GAIN
S3 IN OUT
C13 .022
R17 33K.
MIC.3 J3
T3
C3 .1
R19 2.2M.
R22 12K.
14.6 V.
LO-CUT FILTER
Q3
Q7
C10 .47
R18 470K.
R20 100K.
R21 1.2K.
R23 50K.
CH. 3 GAIN
S4 IN OUT
C14 .022
R24 33K.
J4
GRN.
S5-a
BRIDGE
TERM.
R25 1.2K.
R26 1.2K.
YLW.
S5-b
BRIDGE-TERMINATE SWITCH
R27 33K.
S6-a
R71 100
R72 100
MIC.
LINE
S6-b
R28 33K.
T4
MIC. 4-LINE SWITCH
C4 .1
R31 2.2M.
R34 12K.
14.6V
LO-CUT FILTER
Q4
C11 .47
R30 470K.
Q8
R35 50K.
R32 100K.
R33 1.2K.
CH. 4 GAIN
S7 IN OUT
C15 .022
R36 33K.
MIX. BUS
J9
R37 6.8K.

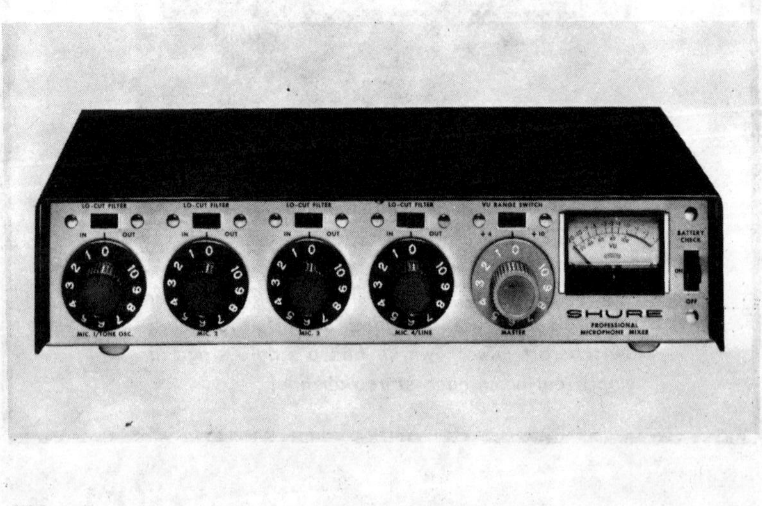
LO-CUT FILTER
IN
OUT
VU RANGE SWITCH
BATTERY CHECK
ON
OFF
MIC. 1/TONE OSC.
MIC. 2
MIC. 3
MIC 4/LINE
MASTER
SHURE
PROFESSIONAL MICROPHONE MIXER

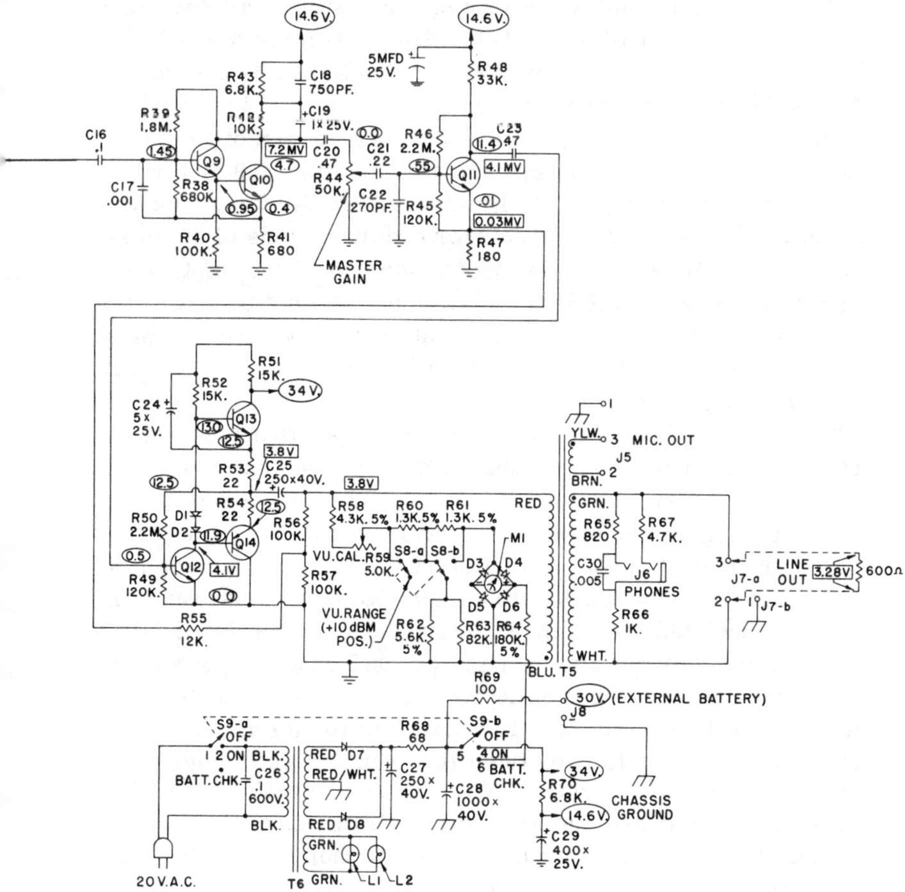

Fig. 9-4. This professional quality M67 mono mixer by Shure offers flexibility as indicated by the schematic, and it can be stacked with a twin for stereo use.

---

price that you're willing to pay. Another decision is whether or not you need a monophonic or a stereo mixer. Most mixers are monophonic, especially the professional jobs, and when you buy a pro mixer you'll need two if you want stereo.

LOW-COST UNITS

There are some inexpensive amplified stereo mixers such as a Lafayette unit which sells for $10. A similar monophonic transistor mike mixer from Lafayette sells for a little under $5. A Switchcraft transistorized stereo mixer is priced at about $25 (see Fig. 9-3). To give a further idea of the price spread, mixers from Bogen are available for $49 and from Shure for $73 (see Fig. 9-4). Shure, by the way, makes several different mixer models for different kinds of applications, and all of them are professional quality. But, once again, you need two for stereo, which might not be a bad idea because of the number of mike inputs that they have. A Sony stereo mixer (see Fig. 9-5) sells for under $30, but it is for low-impedance use only.

You can build a mixer to your own specifications rather easily. Just what you put into it depends on how much you want to spend and how much time you want to devote to the project. Any mixer should have at least two microphone inputs per channel. Ideally, in a stereo mixer each microphone input should be switchable to either channel. This way you could conceivably have three microphone inputs for the left channel and only one for the right channel on a 4-input mixer. For that matter, the switches would permit it to be used monophonically. The input impedance of the microphone circuit for each input is something that you'll have to decide on.

One way to provide good low-noise amplification with high impedance is to use the field-effect transistor (FET). The trouble is that you really can't do this economically with each channel. So an easier method is to use an inexpensive audio transistor with the input circuit adjusted to give a high input resistance. It might be advisable to provide an alternate low-impedance input on at least a couple of microphone jacks for those rare occasions when you must use a low-impedance mike. You can do this with a switch selector at the input (see Fig. 9-6).

Generally, the microphone jacks on the mixer should be standard phone type jacks while the left and right outputs should be phono jacks. The differences between these jacks are described in Chapter 12.

For a mixer to work well and effectively in a variety of situations, it should be battery powered. This also eliminates

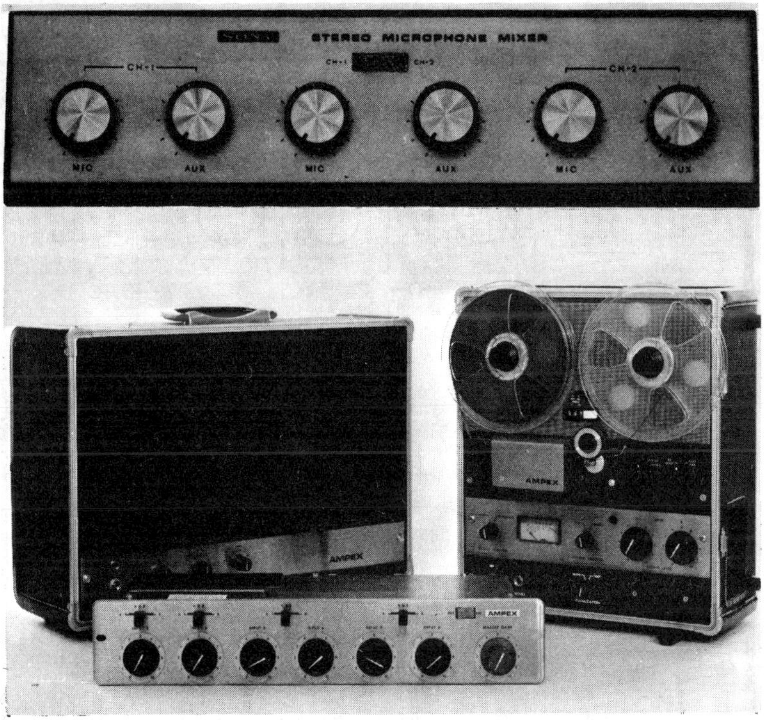

Fig. 9-5. Tagged at less than $25, the Sony MX-6S mixer (bottom) is designed for low-impedance microphones with Japanese plugs. Switchable inputs are featured in the Ampex stereo microphone mixer (top), with three positions for each input: left, right, or middle.

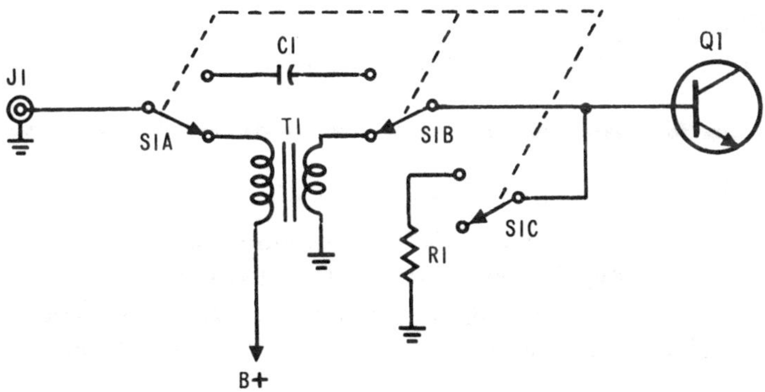

Fig. 9-6. With this circuit, impedance can be switched from high to low for a single input (J1). Switch S1 is shown in the hi-Z position.

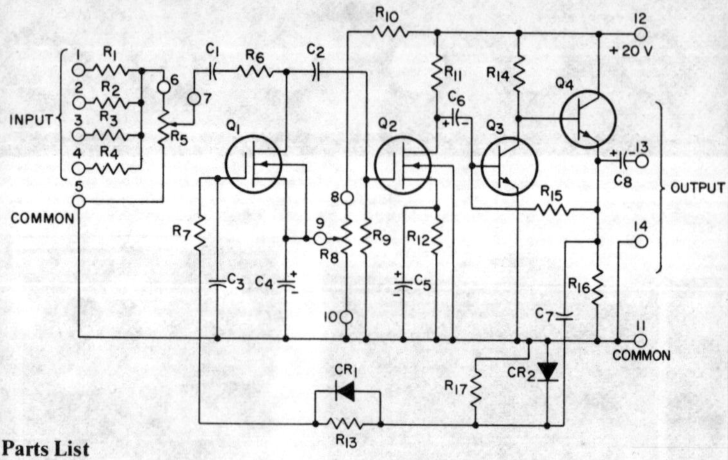

Parts List

$C_1 C_2 C_3 C_7$ = 0.1 microfarad, paper
C_4 = 10 microfarads, 12 volts, electrolytic
C_5 = 15 microfarads, 6 volts, electrolytic
C_6 = 5 microfarads, 25 volts, electrolytic
C_8 = 50 microfarads, 25 volts, electrolytic
CR_1 CR_2 = silicon rectifier, type 1N270
Q_1 Q_2 = MOS field-effect transistor, type 3N128
$Q_3 Q_4$ = transistor, RCA SK3020

$R_1 R_2 R_3 R_4 R_7$ = 100,000 ohms, 1/2 watt, 10%
R_5 = potentiometer, 10,000 ohms, audio taper
R_6 = 180,000 ohms, 1/2 watt, 10%
R_8 = potentiometer, 5000 ohms, straight taper
$R_9 R_{13}$ = 1 megohm, 1/2 watt, 10%
R_{10} = 15,000 ohms, 1/2 watt, 10%
R_{11} = 10,000 ohms, 1/2 watt, 10%
R_{12} = 1500 ohms, 1/2 watt, 10%
R_{14} = 1200 ohms, 1/2 watt, 10%
R_{15} = 100,000 ohms, 1/2 watt, 10%
R_{16} = 470 ohms, 1/2 watt, 10%
R_{17} = 2 megohms, 1/2 watt, 10%

Fig. 9-7. This relatively inexpensive RCA monophonic compressor/amplifier schematic offers four mixer inputs—by R1, R2, R3, and R4.

the need for cumbersome power supplies and the possibility of recording any AC hum. Again, the mixer has to be tailored to your own needs and specifications, so there's a lot of room for variations on the theme. A workable low-noise circuit is shown in Fig. 9-7.

Chapter 10

Remote Controls & VOXs

With the advent of the battery-operated tape recorder came other developments: the remote control and the voice-operated microphone. The VOX (voice-operated mike) is offered as a feature in many Japanese battery portables and can be purchased as a separate accessory to use with any tape recorder with remote control capability.

First, let's take a look at the remote control operating. In most tape recorders, the record function can be turned on—all the circuits are turned on, but the motor isn't running. Remote control is simply an on-off switch in the motor circuit (see Fig. 10-1). This means that while the motor is off, there's still a certain amount of battery drain and the recorder itself will start up with a little bit of wow the instant the re-

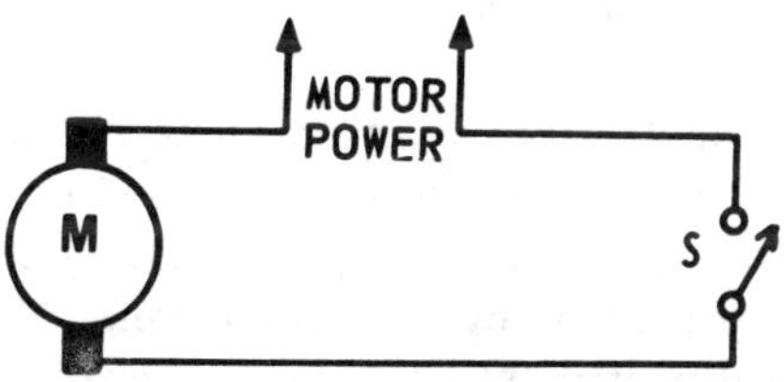

Fig. 10-1. This is the most basic remote-control switch setup using a switch "S" at some location away from the recorder to control drive motor M.

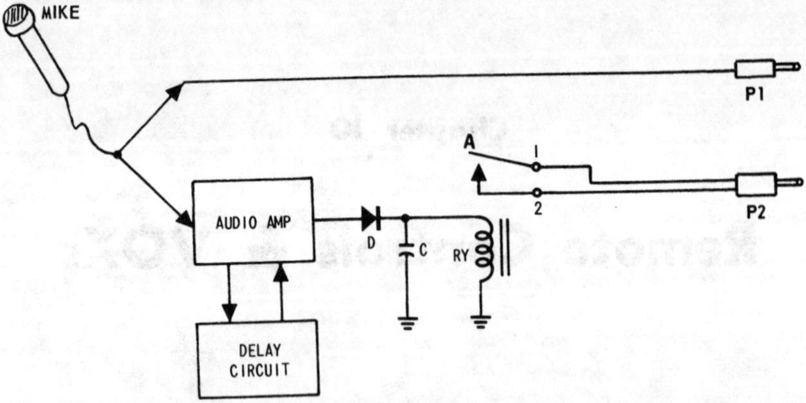

Fig. 10-2. With this basic VOX circuit a signal from the microphone operates relay RY. The mike signal level is amplified and rectified by diode D. The delay circuit prevents the relay from dropping out the instant the sound stops.

mote control switch is flipped on. The wow is to be expected, since it takes a certain length of time (a fraction of a second, usually) to get the tape up to full speed. Any of the normal accessories—foot pedal and other remote type controls—operate exactly the same way. They simply substitute a switch in one form or another for the one that's normally found on the microphone.

VOICE – OPERATED SWITCH

The VOX circuit is simply an extension of the above technique. The switch in this case is comprised of two relay contacts in the VOX circuit (see Fig. 10-2). The relay closes when a voice is picked up by the microphone. The microphone feeds not only directly into the tape recorder but a branch tap-off circuit drives a second audio amplifier of three or four stages which then provides a driving signal for the VOX relay. A rectifier and filter circuit at the amplifier output provides a DC voltage to operate the relay. The relay sensitivity is adjusted with the volume of the amplifier. The amplifier itself also has a time-delay circuit, so that the relay will not open when there is just a pause in the conversation. On some VOX circuits, this time delay is adjustable (see Fig. 10-3).

But as with any remote control device, there is a certain delay in recorder startup and the associated wow in the time limit required for the motor to start. There's also the time necessary for the VOX control to be activated by the voice that is actually controlling the tape recorder. Invariably, you're going to lose the first couple of syllables of any conversation controlled this way. But the beauty of this device is that you get most of the conversation which otherwise wouldn't be possible at all in many situations. The VOX control tape recorder is a very important adjunct to surveillance and clandestine listening devices, especially for telephone taps. There'll be more on this in Chapter 21.

If a tape recorder is left in its "on" position with a remote control switch or VOX control turned off for any period of time, it's quite natural for a "flat" to develop on the rubber pinch roller. Expect to make frequent changes of the rubber pinch roller in such cases. Also, don't rely on such machines for fidelity recordings, since you may have a flat that won't show up until you play back a tape later on with loads of flutter from those flats. It's a good idea to restrict this type of use to one machine.

Almost any tape recorder can be adapted to a VOX circuit, although the easiest are the Japanese battery-operated portables with their remote control jacks (see Fig. 10-4). Here, all you have to do is plug your VOX control into the remote jack and you're in business. Some tape recorders will require some modification. If you have a battery portable that has no remote control outlet, it's a simple matter to break into the motor control circuit and run a pair of wires to a closed-circuit jack—the type used for earphones on transistor radios (see Fig. 10-5). If you use a jack like this, be sure that it's not mounted on any metal part of the chassis, since part of the circuit would then be grounded—a very undesirable feature in this circuit.

During operation, the tape recorder will have to be turned on and the recording level set properly, with the VOX control plugged into the new remote control jack. This same jack can be used for any home-brew remote controls that you care to add to the tape recorder.

With a built-in VOX you're not limited to using just one kind of microphone; you can use the VOX with other inputs, such as a telephone pickup coil or an FM radio monitoring a "bug."

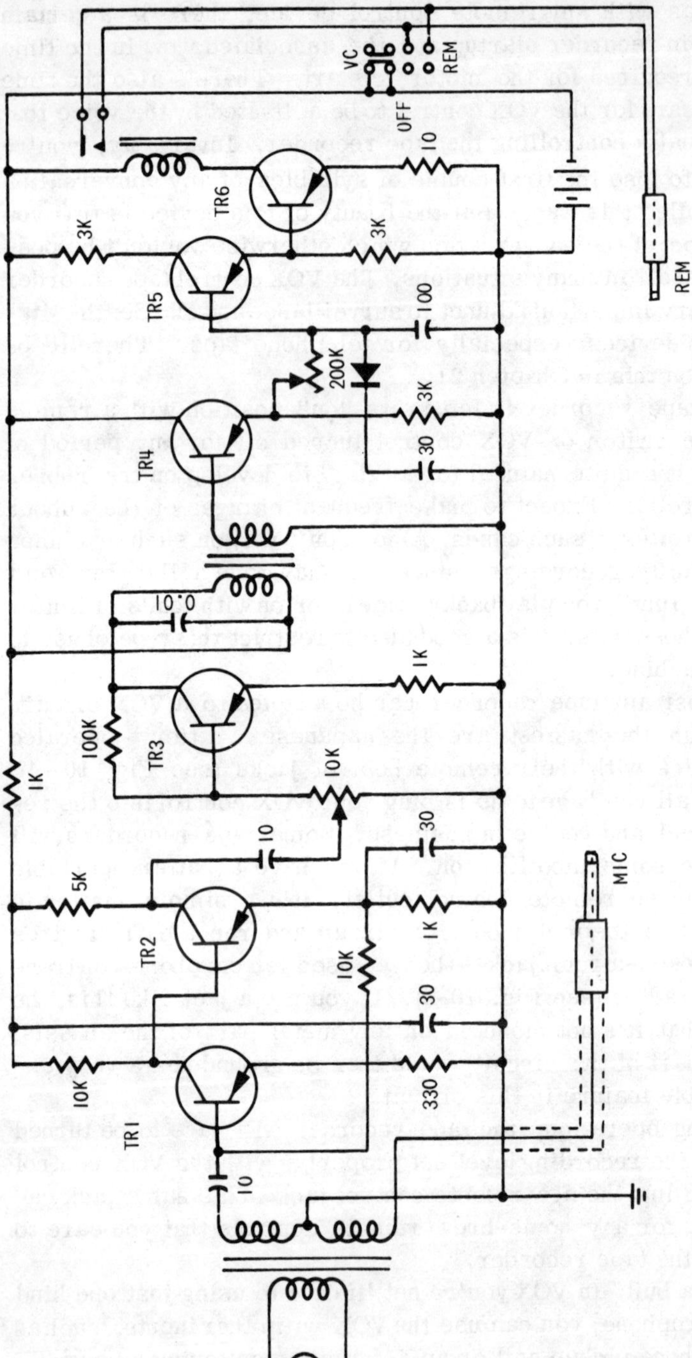

Fig. 10-3. This "typical" VOX circuit is used in Lafayette's voice-actuated microphone.

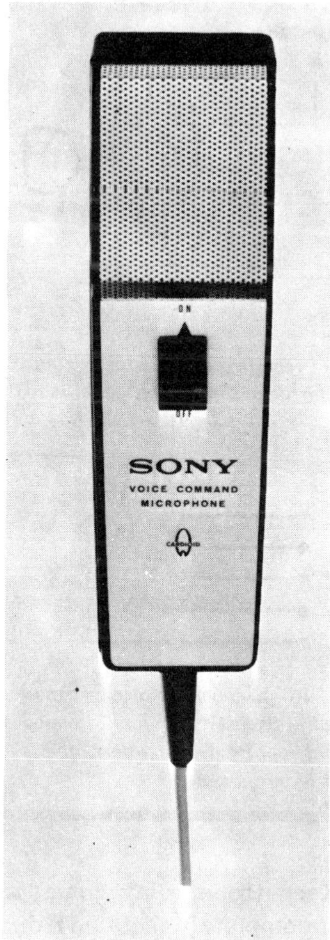

Fig. 10-4. The Sony VOX microphone pictured here is designed for use with a number of different type recorder types.

VOX WITH AC MACHINES

Yes, you can use a VOX control on an AC machine (including vacuum-tube types), just the same way you'd use a switch-type remote control. Just be sure that the VOX you use has no common electrical point between the control circuit and the

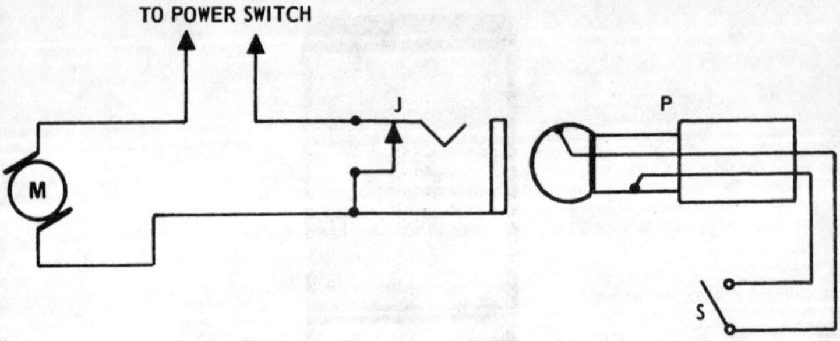

Fig. 10-5. Typical remote-control jack wiring with closed-circuit connection. When no remote-control plug is used the jack completes the motor circuit.

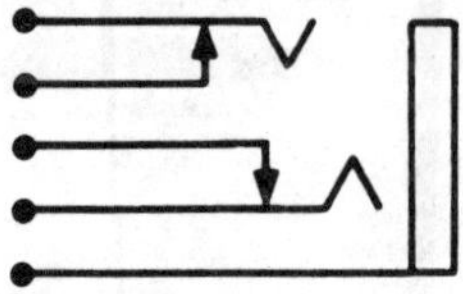

Fig. 10-6. Schematic of a 3-conductor closed - circuit type jack needed for special applications where the shaft might be grounded.

relay contacts. Keep those relay contacts and the plugs and jacks in the circuit completely isolated from any ground at the VOX end and the recorder end of the circuit. A handy way to do this is to use a 3-conductor plug at the recorder—especially if you have to mount the jack on the metal chassis (which automatically puts one circuit at chassis ground potential). This way, you'll still have two conductors left to use for the VOX control circuit (see Fig. 10-6).

Chapter 11

Switching Adds Versatility

If you are a died-in-the-wool tape bug, no doubt you have a pretty decent tape deck wired into your stereo. If so, there'll come a time when you'll want to add a second deck or other recorder to be used as a deck. Adding the second machine gives you lots of flexibility. It lets you duplicate tape, it lets you edit tapes electronically, and gives you freedom to prepare all kinds of specially programmed tapes. But there's a fly in the ointment; adding a second tape recorder to your stereo rig clearly reveals your amplifier's switching limitations.

MAKING THE HOOKUP

When your tape deck is plugged into the tape input and output jacks, the switching arrangement on the preamp is set up so you must flip to the "tape monitor" position on another selector switch when playing back on the tape recorder. Any music source that you have (FM tuner, records, or what have you) will provide a signal at the tape output jacks for recording. But anything played back through the tape input jacks on the monitor position will not provide a signal at the tape output jacks. The reason for this is simple—it prevents feedback.

When you add another recorder, you'll probably take its output and plug it into the "extra" or "auxiliary" jack. Anything you play on this machine will provide an output signal at the "tape output" jack. So now you can easily make a tape-to-tape transfer through the preamp. But what about a transfer in the other direction? Suppose you want to play a tape on machine B and record it on machine A instead of vice versa?

Fig. 11-1. A tangle of cables at the rear of a typical stereo amplifier clearly indicates the need for switching.

Suddenly you find yourself doing just what you had hoped to avoid—pulling plugs, getting tangled up in jumper cables, replugging and hoping you didn't break a connection in those all-too-fragile phono plugs in the process. And when you want to switch back to recording in the opposite direction—another plug-pulling contest (see Fig. 11-1).

If you decide to add still another tape recorder to the collection, such as a cassette or cartridge machine, then you really have problems. At this point, you'll run out of tape inputs in your amplifier and be forced to resort to jumper cables.

There's a simple solution to the problem, and you can put it together for $2 or $3. It's an add-on switch which lets you expand that "extra" stereo pair to as many as four or five stereo pairs. In this particular project (see Fig. 11-2), we've shown four pairs of inputs for three tape recorders and one "extra" pair for such inputs as a TV set, shortwave, Muzak, or whatever other program sources you may want to use.

BUILDING THE STEREO SWITCHBOX

The switch is housed in a small conventional Minibox, which holds a 4-position, 2-pole 4-throw rotary switch. Ten single-hole mounting phono jacks are used for connection points. The two output jacks are mounted at the rear of the box for easy identification, while the other jacks are arranged along the top for easy accessibility. If you feel you need more inputs, don't stop at four pairs. You can expand this to any number of switch-selectable stereo inputs to suit your needs. Just be sure you get a switch that will do the job.

Use shielded cable from the switch's center wipers to the output jacks. The box will act as a shield for the signals—just keep non-shielded wire runs very short (see Fig. 11-3).

Working in the opposite direction can be more of a problem. If you find yourself wanting to make two or more copies of a master tape at a time, chances are you won't get enough signal from your amplifier to drive so many tape recorders. You may want to make a duplicate on open-reel and a cassette tape at the same time. You may want to throw in an extra reel of tape for a friend or a "tapespondent." In any event, something has to be done, and a simple switchbox isn't the entire answer.

BUFFER AMPLIFIER

For one thing, you'll need some kind of buffering action to prevent inter-action among the various tape recorders in-

Fig. 11-2. Switchbox input selector has four stereo input pairs, and one stereo output pair.

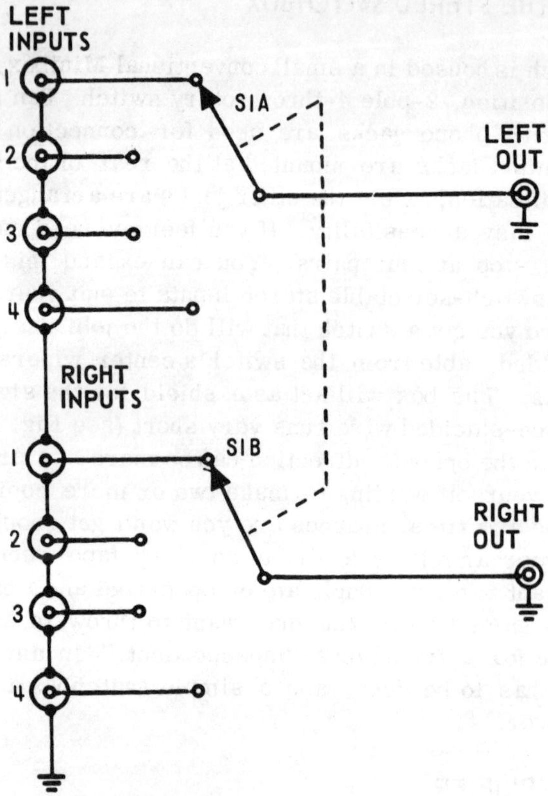

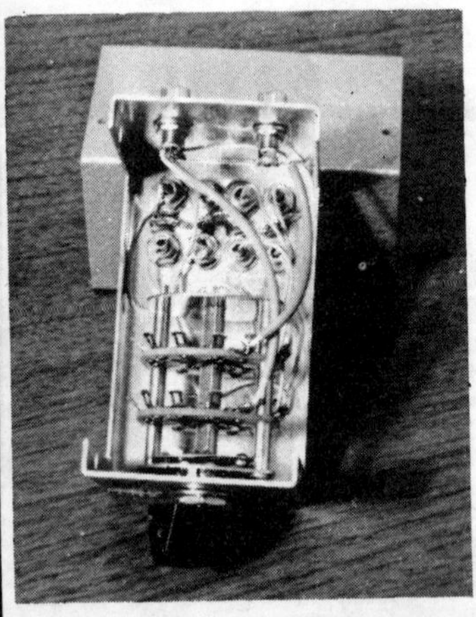

Fig. 11-3. The schematic of the unit here shows how a double-pole 4-throw is wired to provide four stereo feeds to a single recorder input. The Minibox acts as a common ground and shield.

volved. You'll also need something to clean up the signal, since very often there's a lot of background noise and other "junk" present at the preamp's stereo output, disturbances that don't become noticeable until you get down to the two or three recordings made simultaneously.

Another problem arises when you want to be able to switch recorders into the circuit independently of one another. You should be able to switch in one recorder by itself or three recorders at the same time. Here, a simple rotary switch won't do. Instead, the switchbox uses double-pole, double-throw (DPDT) slide switches for each of the three recorders connected to it (see Fig. 11-4).

The signal is cleaned up and buffered by a dual-Darlington integrated circuit amplifier (the same IC amplifier used in some low-level phonograph pickups): The IC is so tiny and lightweight (see Fig. 11-5) that it nestles snugly in the phonograph pickup arm itself. In this circuit layout, it needs some rather hefty outboard components, so you can't exactly hide it in the phono pickup arm.

The circuit (in Fig. 11-6) is an RCA type CA 3036 IC, which is in a 10-pin TO-5 case. The circuit is easy enough to hook up. Connections are made only to the even-numbered pins. You can start off by fanning out the pins and snipping off pins 1, 3, 5, 7, and 9. Pin 6 is the common ground, while pins 4 and 8 are the inputs for the left and right channels, respectively. Output signals for the left and right channels, respectively, are pins 2 and 10. Coupling capacitor C1 is 10

Fig. 11-4. Switchbox with three slide-switch controlled outputs to drive three tape recorders simultaneously. Notice the integrated circuit almost lost from view between the two-terminal lug strips.

Fig. 11-5. A thimblefull of integrated circuits stacked up against a common 9-volt transistor radio battery provides a striking size comparison.

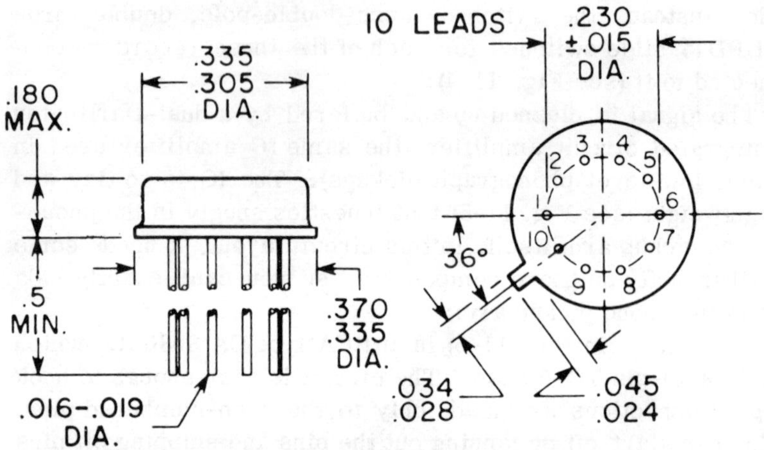

Fig. 11-6. Pin basing diagram of the CA 3036 integrated circuit, bottom view. Only the even-numbered pins are used, so you can snip off the odd-numbered pins, close to the case.

mfd (see Fig. 11-7), both in the left and right channels. The biasing resistor at the input of each channel is 6.8 megohms. Output coupling capacitors are 100 mfd each, and the B-plus connection from pins 2 and 10 are fed through 2.2K resistors. The battery is an ordinary 9-volt transistor-radio type and is hooked up for positive ground.

CLEAN SOUND

When you use this circuit, one of the first things you will notice is the very clean sound that comes out of it. It won't be until then that you realize you've been recording dirty sound in the past. There's a reason for this. Every circuit and every connection in the signal path adds its own little bit of noise, frequency loss, and distortion to the signal. This is

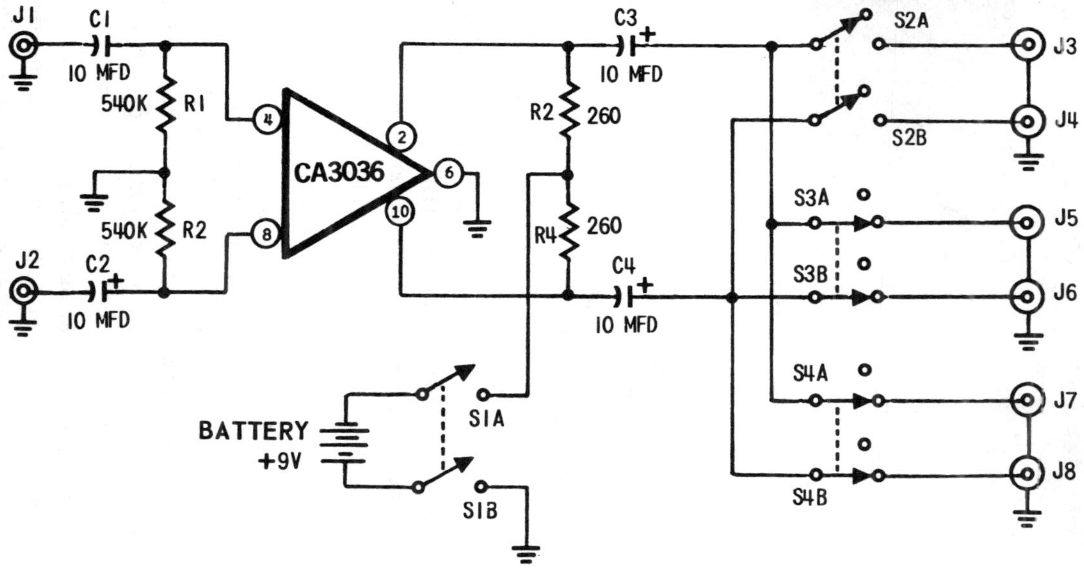

Fig. 11-7. This basic slave driver circuit provides three stereo output pairs from a single input pair.

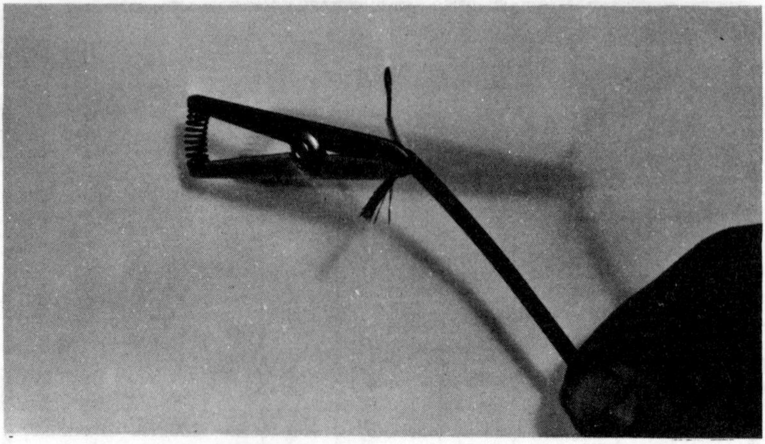

Fig. 11-8. Use a heat-sink tool when you solder and save yourself lots of grief.

why front-panel tape recording connectors are provided on some of the older stereo tuners. This is also why it's an excellent idea to connect directly from one tape recorder to another in doing a tape-to-tape transfer, unless you are absolutely certain of the noise characteristics of your preamplifier.

In any event, it might be a good idea to experiment with short-path recording—recording hookups that eliminate many circuit elements and intermediate connections. Remember, the longer the circuit path, the greater the opportunity for your signal to be degraded.

A word to the wise in hooking up the integrated circuit: It's just as sensitive to heat as any other semiconductor, so treat it accordingly. Use a heat-sink tool on each lead when you solder it (see Fig. 11-8) and leave the heat sink clamped in place long enough to conduct away any residual heat. Use a light touch with the soldering iron—get in and out of there in a split second if possible. And be sure to use a very low-wattage iron that is in good operating condition.

The circuit layout is set up so that all the other connections can be made and soldered first, leaving the integrated circuit to the very end. The IC works like a charm if given half a chance, so take it easy with that iron!

Chapter 12

Plugs, Jacks & Cables

Probably one of the most distressing areas to the tape novice is the somewhat confusing array of various types of connectors and cables that are available. The confusion is understandable; there are certainly plenty of different types to keep tabs on. In some cases, connectors are absent entirely on new equipment (as in the case of quality microphones) and it's up to the hobbyist to install his own plugs, whether he wants to or not?

CABLE TYPES

The way plugs are used and installed depends pretty much on the type of wire or cable they're used with. Understanding cables and how to connect them is important if you want to get consistently good recorded stereo sound. There are four fearful gremlins that you have to beware of in selecting cable: high-frequency losses, hum pickup, impedance mismatch, and capacitance between conductors.

Coaxial cable is a handy solution to all of these problems. This is commonly known as "shielded phono cable" or "shielded coaxial microphone cable" (see Fig. 12-1). Starting from the inside and working toward the outer skin, there's a center conductor made of flexible stranded wire, usually tinned with solder so it has a silver color. The individual strands are tinned so they don't stick together. This wire is surrounded by a layer of polyethylene plastic which is an excellent insulator.

Surrounding the polyethylene is a wire shield, usually made of many tinned strands of wire knitted together to form a basket-weave braid. This shield is the second conductor;

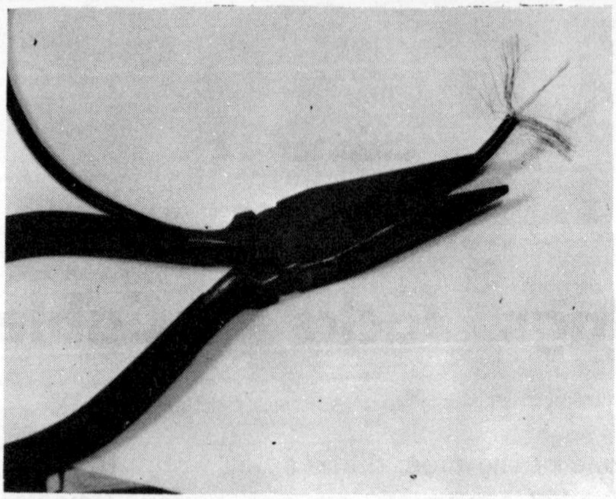

Fig. 12-1. Coax cable consists of a center conductor in a plastic insulating sleeve, surrounded by a shield braid, and then another layer of insulation.

that's why this wire is often called "shielded cable." In some cases the shield layer is simply many strands of wire wrapped around the polyethylene in a spiral wrap with no braiding. The braid gives better structural strength and more positive shielding, but the spiral non-braided shield is easier to work with.

Because of the shielding effect and the unbalance between the center conductor and the braid (the shield is usually at negative ground potential), these cables tend to retain high frequencies which would be lost with ordinary lamp cord or other non-shelded wire. Also, because the shield is at ground potential, it protects the circuit from interference by stray AC fields which would otherwise induce hum in the line.

Impedance matching is another problem. Generally, coaxial cables have a rather low impedance, and using them in the circuit won't cause any serious mismatches as long as they're kept reasonably short. A good rule-of-thumb to follow is "keep the cables as short as possible." This not only holds the cable impedance down to reasonable proportions, it reduces the signal's tendency to leak off those high frequencies. Cable length of 20 to 30 inches is appropriate. Capacitance —which could affect signal properties—is also minimal with coax. There's some capacitance buildup in such cable, but not enough to affect audio frequencies.

DON'T ALWAYS USE COAX

There's one very important place where coax should not be used—between the amplifier and the speakers. This is one place where ordinary lamp cord (zip cord) or darn near anything else that's handy can be used. In some cases, even flat TV antenna twin-lead is preferable, especially where you want to run speaker wires under the living-room carpet without noticeable bulges. But keep these runs of TV wire short: the line impedance for TV twin-lead is high and can cause a mismatch if too much of it is used.

The best speaker wire to use is a small-diameter zip cord that looks like miniature lamp cord. It's sold as "speaker wire" by most radio parts distributors for about 2¢ a foot. Pick up a 100-foot roll of it; it'll get used up before you know it. This wire has a transparent plastic coating so you can see the wires inside. The wires are "color-coded"—one is raw copper, the other is tinned so it has a silver appearance. This color-coding is crucial for proper speaker plasing (see Fig. 12-2).

THE PHONO PLUG

The American high-fidelity industry is cursed with a standard plug introduced by RCA in the 1930's. It's known simply as

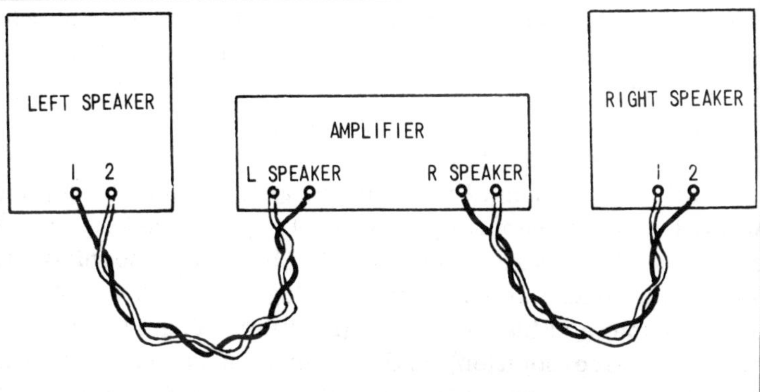

Fig. 12-2. Phasing speakers with color-coded speaker wire—one wire is copper and the other is silver-colored. Just keep those wire connections consistent and the speakers will be properly phased.

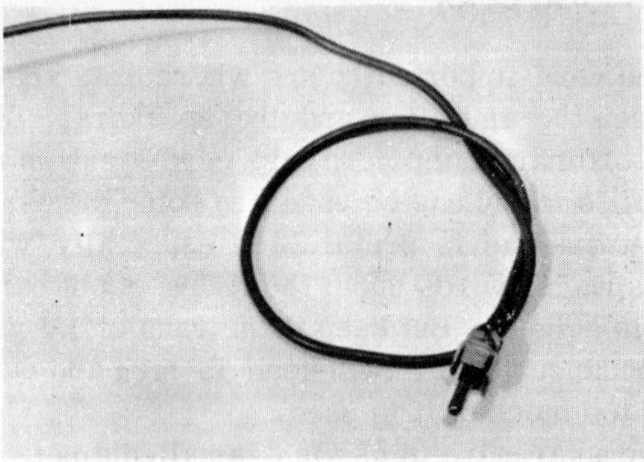

Fig. 12-3. The basic RCA-type phono plug has become a universal standard in the U.S.

a "phono plug" today (see Fig. 12-3), but you also may hear it called an "RCA plug" from time to time. The plug is a curse because it's frequently hard to solder properly, and in its bare form it offers no finger grip for unplugging. Very often, because of this lack of finger grip, too much strain is put on the cable, which then develops an intermittent or broken connection. This problem is a common one, and whenever there's an intermittent connection, noise, or short (causing hum or lack of signal), the phono plug is the first place to look. To test it for intermittent problems, simply flex the cable near the plug with the equipment turned on. If the trouble appears and disappears with the flexing, look no further. That's the gremlin right there. Snip the cable off and resolder it to a brand-new phono plug (see Fig. 12-4).

There are several ways to lick the phono-plug syndrome. One is to buy cables already made up with molded-on plugs. Molded-on plugs generally have finger-grips, making unplugging less of a chore. You can also buy more sophisticated phono plugs that give you a varying degree of added protection for the cable and lots of finger-grip area. There's no simple or inexpensive solution, and the ready-made cable isn't always the best answer since you'll have specific cable length requirements of your own. If you can make up your own cables and then avoid lots of unplugging, you'll be money ahead. One way to avoid frequent unplugging and replugging is to use one or both of the selector switchboxes described in Chapter 11.

THE PHONO JACK

From time to time, you may have to install a phono jack, as in building the Chapter 11 switchbox projects. There are two basic types of phono jacks—the common, conventional metal and phenolic board composite and the single-hole jack that's

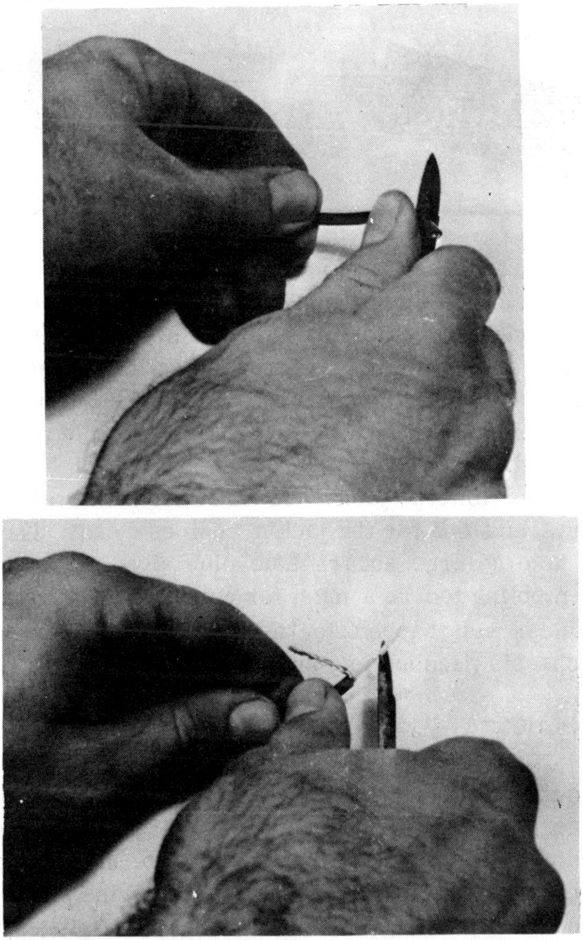

Fig. 12-4. To start stripping coax, slit and remove about one inch of the outer insulation (A). Bunch, unweave, and pull the shield braid to one side and then strip the center conductor about halfway (B).

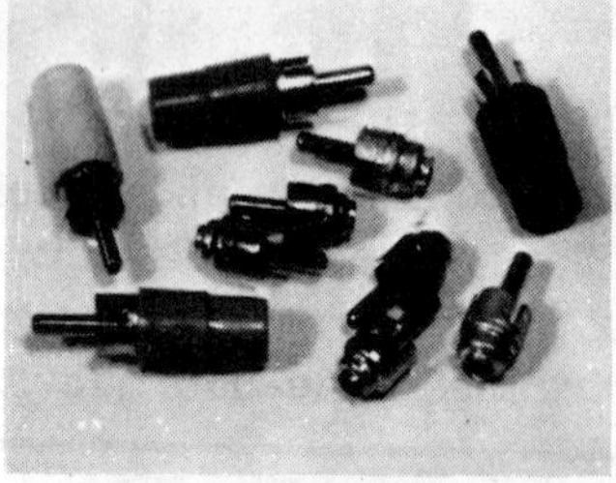

Fig. 12-5. A few possible variations on the RCA-type phono plug (A) and the phono jack (B).

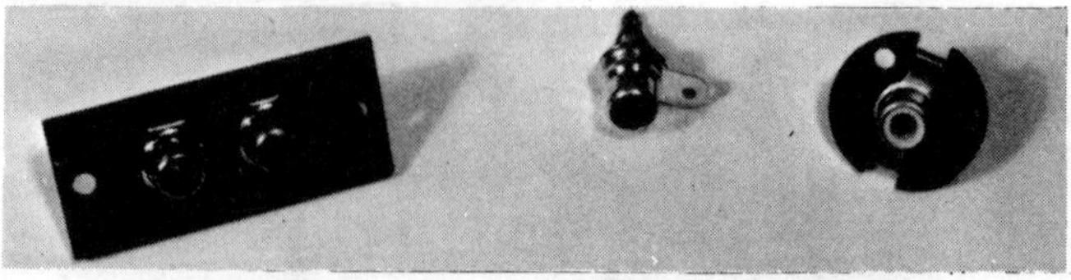

a dandy from the installation viewpoint, even though it's more expensive than the combination type. The single-hole jack requires just one 1/4-inch hole in the chassis. It's held in place with a nut on the inside that screws onto its threaded collar (see Fig. 12-5). These jacks are listed in the Lafayette catalog for 19¢ each, or 10 for $1.70. Buy a couple of sacksful; you'll need them.

The combination type jack takes three holes—two for mounting screws and one for the jack's business end. The double combos take a large square hole plus two mounting holes. Here, a nibbling tool is a must for cutting that large, square hole, unless you happen to have some large, expensive, square chassis punches in your toolbox.

PHONE PLUGS

Lots more sensible in basic design, the phone plug is larger and more costly than the phono plug. It has a large, solid barrel which gives you an excellent grip for unplugging. The phone plug resembles (but isn't the same as) the plugs used in a telephone switchboard (see Fig. 12-6). These plugs are often still referred to by their old military designation—the PL-55 2-conductor phone plug and the skinnier PL-68 3-conductor plug used in switchboards. The two plugs look very much alike, so don't confuse them if you see a surplus bargain somewhere.

When connecting a phone plug to a cable, be sure to remove the bakellite sleeve and feed the cable through the sleeve <u>first</u> (Fig. 12-7). If you forget to slip the sleeve on the wire before soldering, you'll discover that there's no way to do it afterward, and you probably won't make that mistake again!

Next step, after sliding the bakellite sleeve onto the cable, is to solder the shield to the long metal strip that forms the frame of the plug. This is the grounded side, usually with a small slot or hole to accept the braid. Trim off about half of the shield's wires before soldering, otherwise the shield will be too thick to handle. Also, while soldering, grip the braid near its root with a pair of long-nose pliers to conduct away soldering-iron heat. Do this while the soldering iron is applied to the braid and hold the pliers there for a few seconds of cooling time afterward. Otherwise, the soldering heat

Fig. 12-6. Similar in appearance, the telephone-type PL-68 plug (left) has a narrower barrel than the PL-55 or phone plug at right.

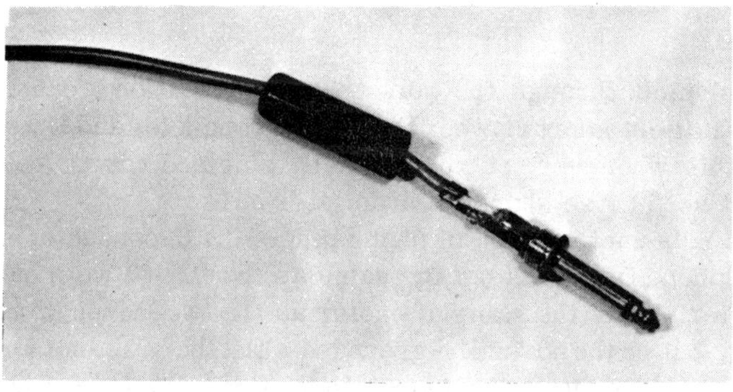

Fig. 12-7. Wiring a two-conductor phone plug is relatively easy; just remember to slip that sleeve over the wire before you start soldering!

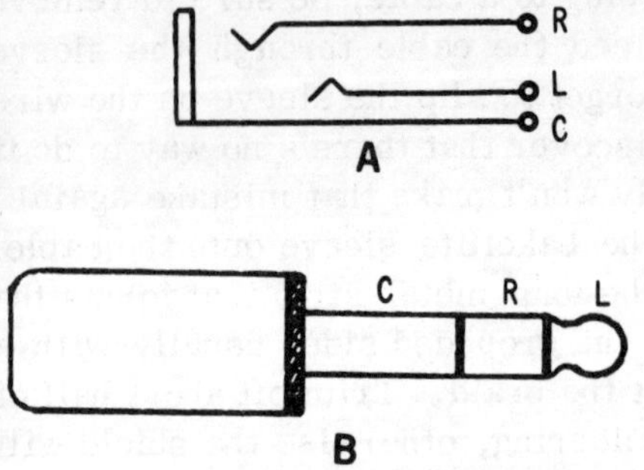

Fig. 12-8. Wiring of a three-conductor phone plug, commonly used for stereo headphones. It has the same size barrel as the two-conductor phone plug.

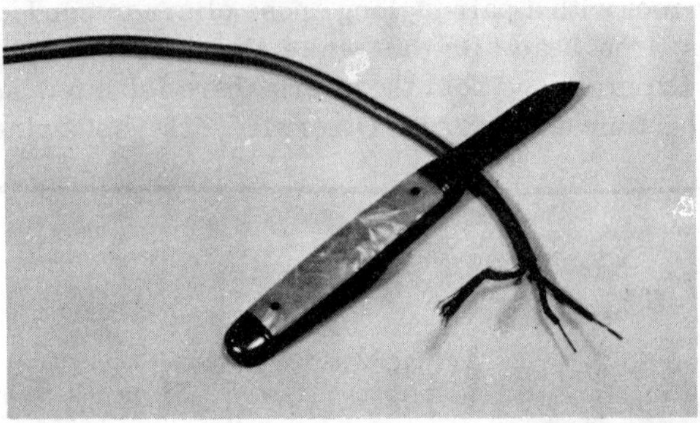

Fig. 12-9. Two-conductor shielded wire commonly used in balanced microphone lines and in some stereo wiring installation with a common ground.

might melt through the polyethylene insulation, causing an intermittent short between the center conductor and the shield. Do this whenever you're soldering shielded wire. You sure don't need those shorts to undo later on!

There's another type of phone plug—the 3-conductor stereo headphone type. It's not the same as the PL-68 we mentioned earlier. It is the same diameter as the two-conductor phone plug, but on the normally-grounded shaft there is another conductor added near the indentation at the tip (see Fig. 12-8). This plug was devised mainly for stereo headphones, but it is finding lots of other uses in stereo equipment. It does the job of two plugs and jacks when double cables are normally used.

A 3-conductor plug uses a 2-wire shielded cable. This is not a coaxial cable, since its layers aren't concentric (they're not around the same axis). This cable contains two parallel wires surrounded by a shield (see Fig. 12-9). The shield is grounded, as with coax cables, but this ground is common to both circuits—which is usually the case in stereo equipment. This type of cable has many of the same advantages as coaxial cable, although its upper frequency limit may not be as good. Still, the upper limit is well beyond the audio range, so it's fine for stereo uses. However, there may be some capacitive coupling between the two inner conductors, causing crosstalk between stereo channels. Because of this, keep cable runs short (except for stereo headphones) and avoid using this type of cable in critical circuits. Use the old-fashioned one-cable-per-channel approach instead.

JAPENESE PLUGS

Japanese tape recorders generally use a miniature version of the phone plug. This size has become fairly standardized and is the same size plug used on most transistor radio earphones. These plugs and jacks look exactly like their larger counterparts; they're just scaled down in size (see Fig. 12-10).

There's a variation on phone jacks that you'll have to watch out for. This is a specialized version known as a "closed-circuit" or "shorting-type" jack, which automatically shuts off the transistor radio's loudspeaker when you plug in the earphone (see Fig. 12-11). The jack is a 2-conductor type generally, but has three solder lugs on it. One of the jack's circuits acts as a closed switch when no plug is in and an open switch when the plug is inserted (see Fig. 12-12). Some 3-

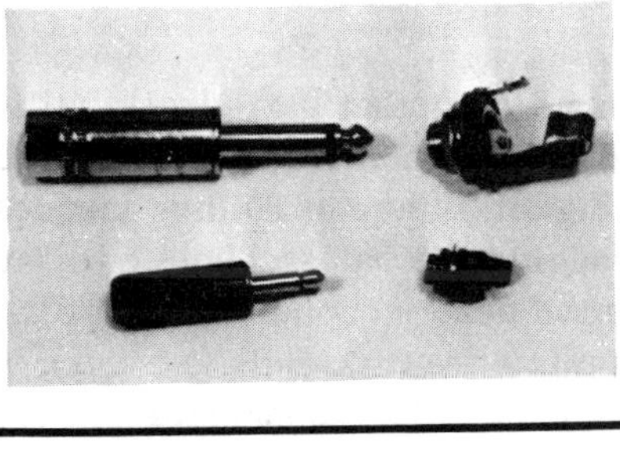

Fig. 12-10. Pictured here is a standard phone plug and jack (top) and the tinier Japanese version.

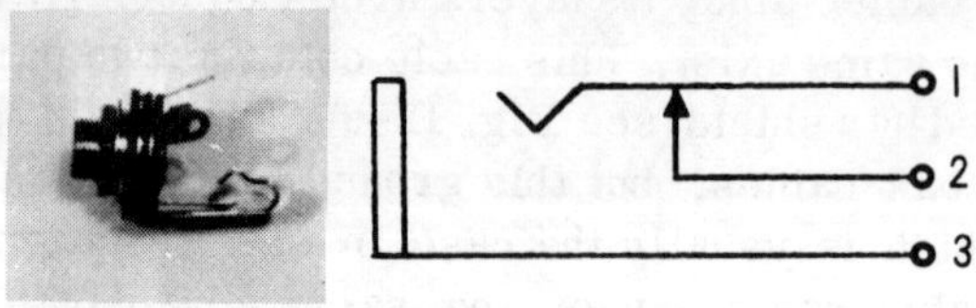

Fig. 12-11. A two-conductor closed-circuit jack opens a normally-closed circuit when a plug is inserted.

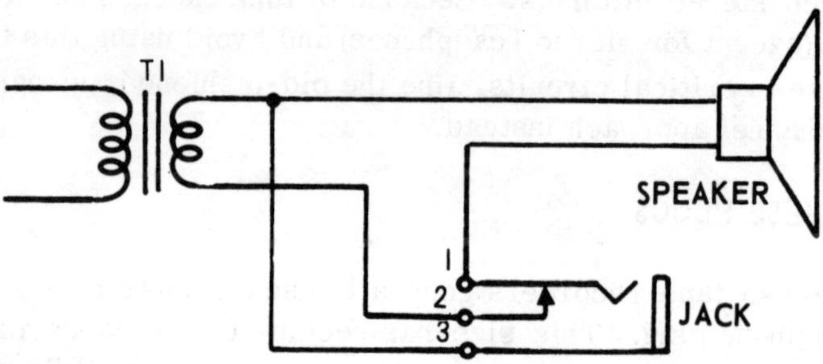

Fig. 12-12. In this schematic a closed circuit jack is used to shut off the speaker when an earphone plug is inserted.

conductor jacks also have this closed-circuit feature. Generally, jacks without this switching element are called "open-circuit" jacks. You can use a closed-circuit jack as an open-circuit jack simply by not connecting its third solder tab.

SCREW-ON PLUGS

One fairly popular microphone plug that's gradually falling into disfavor is the "single-button" contact type (see Fig. 12-13). This connector has the big advantage of holding fast despite plenty of vibration and tugging. It has a couple of disadvantages, too: the single-button center contact can get dirty, making an intermittent contact; screw threads get worn or stripped, making it difficult to use. This connector is popular mainly for amateur radio and communications equipment. You'll rarely find it on a tape recorder, although the now-discontinued EICO tape recorder was one notable exception. The microphone input on most tape recorders is a phone plug. The line input and output for direct connection to the stereo amplifier are made via phono plugs.

As with most connectors, there are numerous adapters to convert it to and from phone plugs and jacks. There are also adapters designed to and from phone and phono, and pre-assembled cable assemblies come in a variety of plug-and-jack combinations. The most useful single type of cable (oftentimes called "jumper" cable) has a phono plug at each end. For some equipment, you'll need a pair of cables with a phono plug on one end and a phone plug on the other. And you may need variations of these.

EUROPEAN PLUGS

Another plug type becoming increasingly well known is the Din plug used on European equipment (see Fig. 12-14). There are two basic plugs—the 3-pin and the 5-pin types. The 3-pin is used for microphone and line inputs and for line outputs.

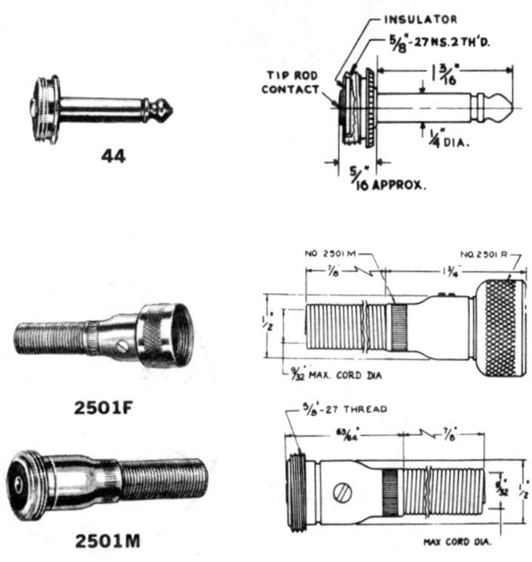

Fig. 12-13. These Switchcraft microphone connectors are often used on communications equipment but rarely on tape recording equipment.

Fig. 12-14. *Typical European or DIN plugs. The plugs at (A) are used with the Norelco Carry-Corder series.*

A

B

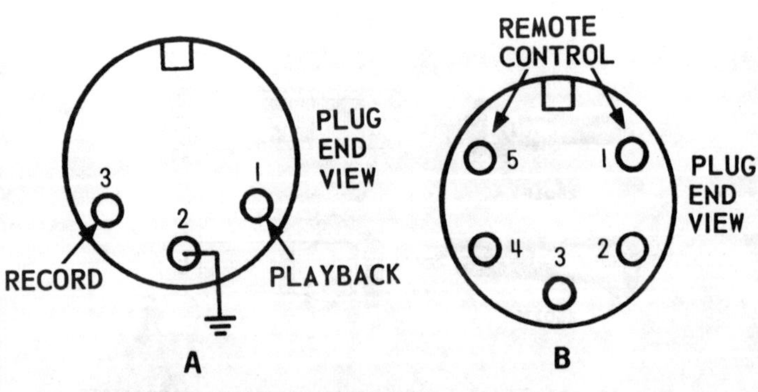

Fig. 12-15. Wiring diagram of DIN plugs used in the Norelco Carry-Corder and Wollensak 4100 and 4200 series.

In some machines, the 3-pin type is used for a stereo pair of inputs, and another one for a stereo pair of outputs. The 5-pin plug is generally used for remote control switches and for battery chargers or AC adapters. The wide outer lip of the 5-pin plug also opens a European version of the closed-circuit jack to activate the remote-control switch plugged into the socket. Since this plug has several different functions, the battery charger plug has the two remote-control pins shorted together.

If a remote control or footswitch is plugged into the tape recorder, the plug has another 5-pin jack dangling from it for plugging in the battery charger or whatever else is handy. On the "dangling" jack the remote control pins aren't connected so nothing will affect the motor switching.

If you want to add an accessory of your own, you'll have to buy the Switchcraft European-type audio connectors. They're rather expensive (compared to other plugs), so don't go overboards with ideas. The 3-pin type sells for 66¢ and the 5-pin list for 78¢ in the Lafayette catalog. To use one of these plugs, remove the screw in the barrel and remove the metal barrel, leaving the rubber sleeve in place. Slip the rubber sleeve and barrel over the cable before you start stripping wires. Finding the right pins is strictly trial-and-error, depending on the tape recorder. Fig. 12-15 shows the pin connections for the Norelco Carry-Corder cassette portable. Many of these connections will be the same for other European machines.

By the way, you may hear plugs and jacks referred to as "male" and "female." The plug is the "male" and the jack or receptacle is the "female" for obvious reasons. If you find it necessary to use these very descriptive terms in a radio parts store, don't blush; it's quite acceptable terminology and is a sure way of telling the boys from the girls.

Chapter 13

Integration With a Stereo System

One of the very nice features of a tape recorder is its versatility. You can use it on the go, at parties, or in your living room piped through your big stereo rig. And, hooked into the big rig in the living room, it can save lots of dollars for you by recording music from borrowed records and from the FM stereo.

If you have a stereo phonograph or console combination of some kind, you probably spend a certain amount of money on records every year. If you have an FM radio built into that stereo rig, you have a ready-made source for building a very low-cost music library. Anything coming in on the radio can be recorded on tape for a fraction of the cost of the original records.

The best way to make a tape recording from a record player, FM radio, or TV set is with a direct electrical connection. The worst way to do it is by putting a microphone in front of the loudspeaker. Microphones supplied with tape recorders generally aren't very good quality; the speakers may not be either, and you'll have all kinds of problems with mismatched acoustics, room noise, and just poor sound fidelity in general.

If you have one of the more complex stereo consoles, it may already have connections on the back for adding a tape recorder. Generally, these will be labeled "tape output" and "tape" or "auxiliary input." You'll need appropriate cables to plug into these jacks. Usually, they are standard phono jacks; the tape recorder's connectors will be either phono or phone jacks.

You can buy ready-made cables for making these connections

at most hi-fi showrooms and radio parts distributors. Just be sure the cable is long enough to reach from the recorder to the equipment you're connecting it to. The jack marked "tape output" on the stereo is connected to the "line input" or "pre-amp input" on the tape recorder. If this is a stereo setup, there'll be two jacks for each input and output, and you'll need two cables—one for each channel. For playback, you can use the tape recorder's own speakers, or if you have pretty good speakers on your stereo equipment, you can run another cable pair from the "line output" on the recorder to the "auxiliary" or "tape input" jacks on your console.

ADDING CONNECTORS

What if you don't have these handy jacks on your stereo? With a little cutting and soldering, you can add the proper connectors to your equipment in one evening. Let's take the

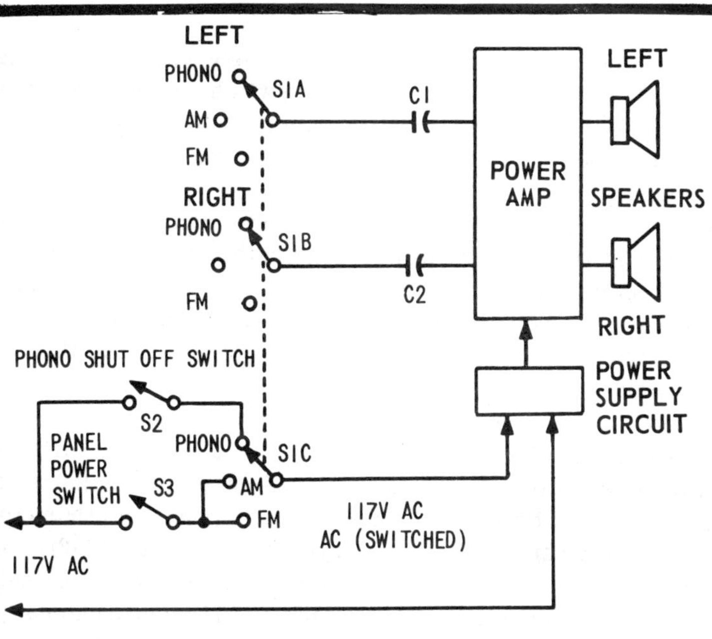

Fig. 13-1. Typical selector switch setup for a console radio-phono. Notice that the power switch is off when the selector is in the phono position; AC is handled by the switch in the record player itself.

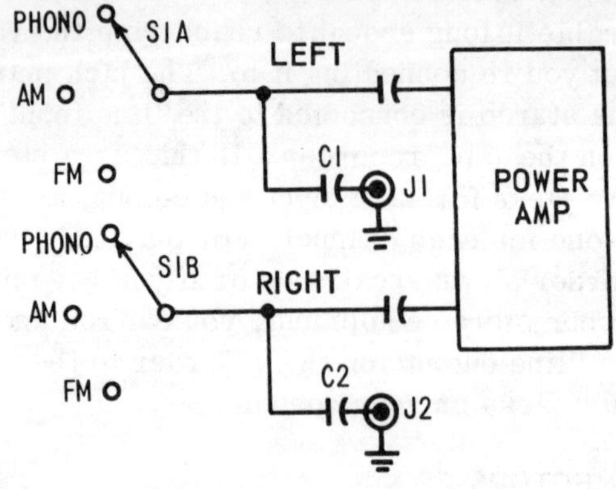

Fig. 13-2. Tape output jacks are fairly simple to add to the console. Capacitors C1 and C2 should be .02 mfd, while jacks J1 and J2 should be standard RCA-type phono jacks.

console-type unit first. If it is a phonograph and a radio, there will be a phono/radio selector switch somewhere on the control panel. Locate the center wiper of the switch for each stereo channel—a terminal that's always making contact with the rotating blade on the wafer (see Fig. 13-1). The wire from this contact goes to the grid of the first amplifier tube or to the base of the first amplifier transistor in the circuit. It may be interrupted by a coupling capacitor (about .02 mfd to 10 mfd for transistor circuits). There will also be a switch section that handles 117 volts AC for automatic shutoff in the phono mode. Don't touch this section.

Run a wire connection from this center wiper terminal through a .02-mfd capacitor (for tube-type consoles—1 mfd for transistor units) to the center terminal of the output jack, which you can mount at the rear of the cabinet or at any other convenient spot where it won't be an eyesore (see Fig. 13-2). For stereo units, of course, you'll have to run a cable from each identical switch section. Be sure to use shielded phono cable, connecting the center conductor to these signal points. The shield braid should go to a ground connection (the chassis). Hooked up this way, no matter what is being played on your stereo—records, FM, or AM radio—you'll be able to record it directly.

A less satisfactory but much easier method is to attach leads to the speaker connections. If your speakers don't have readily accessible terminals and are the plug-in type, you can make a "Y" connecting cable, or buy one ready-made for each channel (see Fig. 13-3). The speaker will plug into one jack, the tape recorder into the other, and the tail of the "Y" will plug into the unit's speaker jack.

If the speaker terminals are exposed, a cable with a pair of alligator clips at one end can be attached to these terminals very easily (see Fig. 13-4). The other end will have an appropriate plug for use with the recorder. With this type of setup, remember that changing the volume control setting on the stereo will change the signal level going to the tape recorder. Use a fairly low volume level at the speakers for best reproduction. For playback, again using alligator clips, the stereo speakers can be hooked into the tape recorder's "extension speaker" connections.

ADAPTING FOR PLAYBACK

A permanently-wired playback setup at the console is preferable, but again requires some cutting and soldering. In this case, you'll also have to add a switch to the unit—a double-pole double-throw (DPDT) rocker, toggle, slide or rotary—whichever is convenient.

Mount the pair of jacks and connect their center connectors to one outside pair of contacts on the new switch. The other outside set of switch contacts goes back to that center wiper on the selector switch, or just to the tape output jacks you've

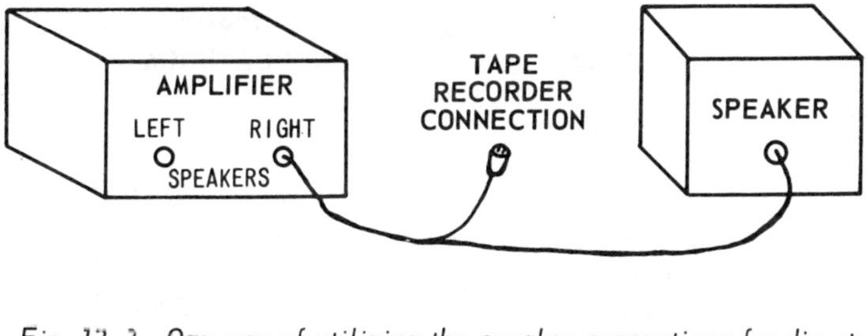

Fig. 13-3. One way of utilizing the speaker connections for direct recording is to install a "Y" connector in the speaker line.

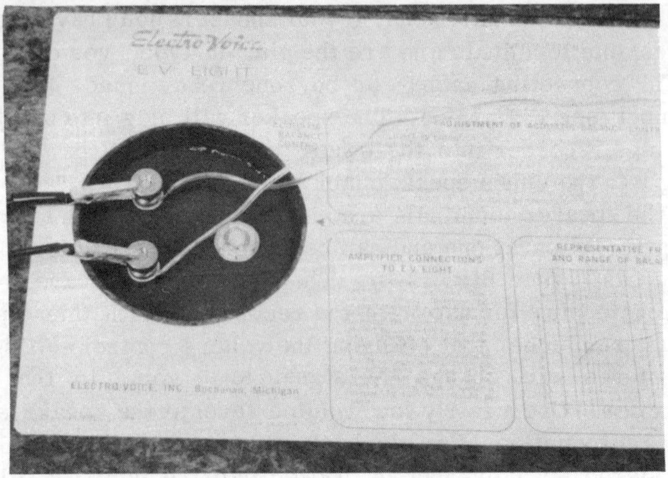

Fig. 13-4. With alligator clips you can "lift" the signal off the loudspeaker terminals.

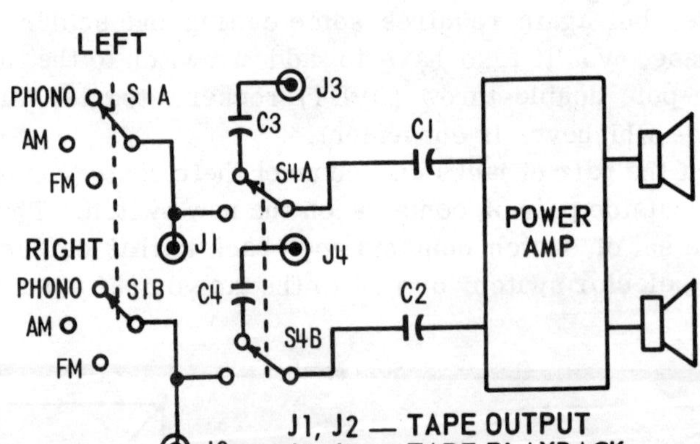

Fig. 13-5. Schematic showing how to add tape playback jacks to the console switching circuit. Jacks J1 and J2 are tape output jacks; J3 and J4 are tape playback jacks. C1 and C2 are already in the existing circuit. C3 and C4 are .02 mfd capacitors in the playback circuit.

already installed. To do this, you'll have to break the signal path from the selector switch to the equipment's amplifier (see Fig. 13-5). The wires from the amplifier itself go to the center wiper contacts of the new DPDT switch.

With this arrangement, the new selector switch has two positions "radio" and "tape." Simply flipping the new switch to "tape" will cut off the radio and switch in the tape recorder as the music source. In the "phono" position, some consoles will shut off unless the phono motor is turned on.

COMPONENT SYSTEMS

If your stereo rig is made up of components, then you're home free. The tape recorder hookup is simply a matter of running a pair of shielded cables between the tape machine's inputs and a pair between the outputs and the jacks provided on your amplifier or preamp (see Fig. 13-6). Many amplifiers even give you an option of how to do this.

First of all, there's the pair of jacks marked "tape output." With cables plugged into this pair and into the recorder's "line input" or "mike input" (if there's no line input), you can record anything that you're listening to on your stereo, be it

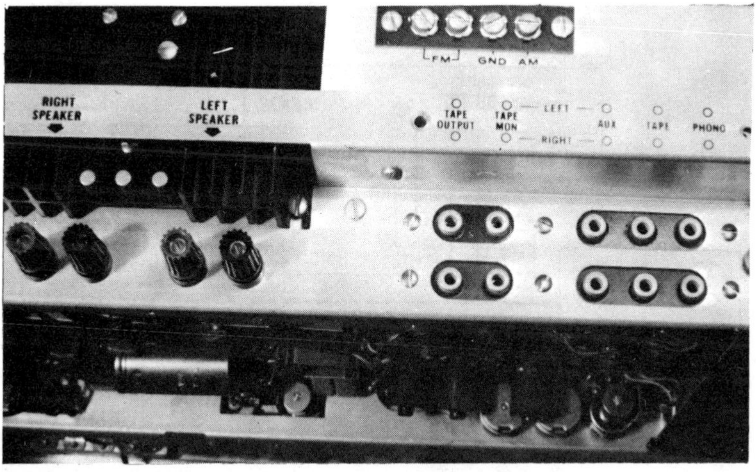

Fig. 13-6. Jack lineup on the back of the Heathkit AR-15 stereo receiver. These jacks make it easy to add a tape recorder.

Fig. 13-7. Side-by-side hookup of two tape decks—the Heathkit AD-16 (left) and Ampex 2150 (right)—makes tape-to-tape transfers and electronic editing feasible.

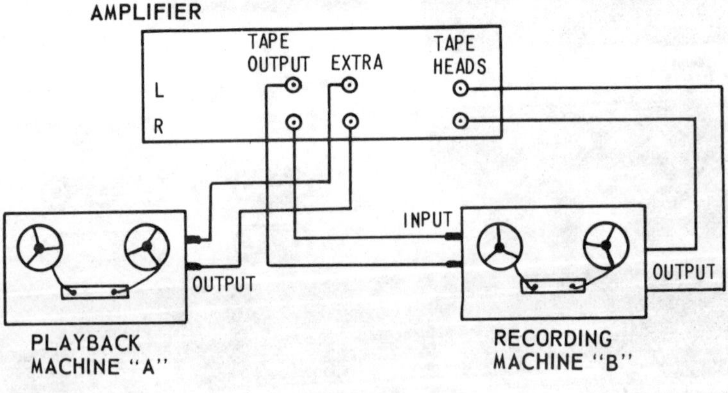

Fig. 13-8. Two tape-deck hookup using a conventional component-type stereo amplifier. Tape machine "A" is used to play the "master" and machine "B" records the duplicate.

records, FM stereo, AM, or another tape recorder. The amplifier may also have a pair of jacks marked "extra" or "auxiliary." Run cables from this pair to the tape recorder's "output" or "line output" jacks. If the recorder doesn't have a similar output, you may have to use its "ext. speaker" jacks. If you're already using the auxiliary input on your amplifier for something else, and it doesn't have separate tape input jacks, then you'll have to unplug and replug each time, or build yourself the input expander switchbox described in Chapter 11.

If your amplifier has another pair of input jacks marked "tape" or "tape heads," you have a bonus in some extra versatility. By plugging into this pair of jacks, you leave the "extra" input free for other equipment. Usually there's a separate "tape monitor" switch on the front panel which cuts in the tape recorder for playback. With this arrangement, you can hook up two tape recorders through your amplifier and make tape-to-tape transfers for duplicating or electronic editing.

To do this, select one machine for recording and one for playback (see Fig. 13-7). The playback machine is plugged into the "extra" inputs and the recording machine's playback goes through the "tape heads" input (see Fig. 13-8). The playback machine's signal will be available at the "tape output" jacks when the selector switch is in the "extra" position, and these jacks should be cabled to the recording machine's "line input" jacks. For test playback of the new tape, switch the amplifier into "tape monitor" position.

All this is fine if you're going through the stereo rig during playback and duplication. But it's always better to make a direct connection between tape recorders for duplication—this shortens the signal path length, the sound will come out cleaner and crisper, with a lesser tendency to lose high frequencies. If your stereo amplifier is really top-quality, then it may not add any coloration or noise of its own. But still, keep those jumper cables short!

Chapter 14

Recording From a Telephone

At some time or another, just about every tape recorder owner wants to record telephone conversations. You've probably seen this done on television, you've heard about tape-recorded phone calls produced as evidence in court cases, you've heard of wire taps (see Chapter 21), and have probably thought of a few useful ways that such a telephone setup could operate for you.

Trouble is that with today's garden-variety transistor tape recorders, taping from the telephone can be very difficult. Few tape recorder manufacturers offer a good, well-matched telephone pickup coil for their equipment. Even if they do, chances are it is grossly overpriced. Lower cost pickups, ranging from $1.25 on up to $3.00 or so, are available in a variety of forms, shapes, and sizes in just about any electronics parts store, but they don't work very well with transistor tape recorders.

First of all, there's the problem of impedance matching. This wasn't a problem back in the days of tube-type tape recorders, since the impedance match was automatic. Vacuum tubes are high-impedance devices, and so are most telephone pickup coils. Also, the first couple of stages of amplification in the tube-type tape recorder give plenty of extra gain which is needed for this. Not so with transistor tape recorders. Many of them operate with only a marginal amount of gain in the first few stages. This may crop up as a problem also in recording from a low-level phonograph cartridge or from another tape recorder with a low-level output. With telephone pickups, it's crucial, because you can crank the gain up all the way on the recording level and still get nothing that's usable or recognizable.

There are several ways to solve this problem. You can build a simple one- or two-transistor preamplifier with appropriate matching at the input, you can go the modular route using prepackaged modules, or you can try to get a tape recorder with a well-matched telephone pickup. In this last case, you'd have to do quite a bit of shopping around because very few battery-operated tape recorders fill this bill.

The many shapes of telephone pickup coils employ basically the same operating principle. All contain an induction coil, which picks up some of the magnetic field radiated by the telephone and transmits this feeble signal to the tape recorder. How well the pickup coil works depends on its design, where it's placed on the telephone, the design of the telephone, and how much amplification the tape recorder has, and, of course, impedance matching.

PICKUP TUBES

The familiar suction-cup pickup is designed to attach to the telephone handset (see Fig. 14-1). The flat bakellite-encased coil is made for desk-type telephones and is designed to fit under the body of the phone (see Fig. 14-2). These particular pickups are a little bit touchier since you have to find the most sensitive position under the phone's coil block.

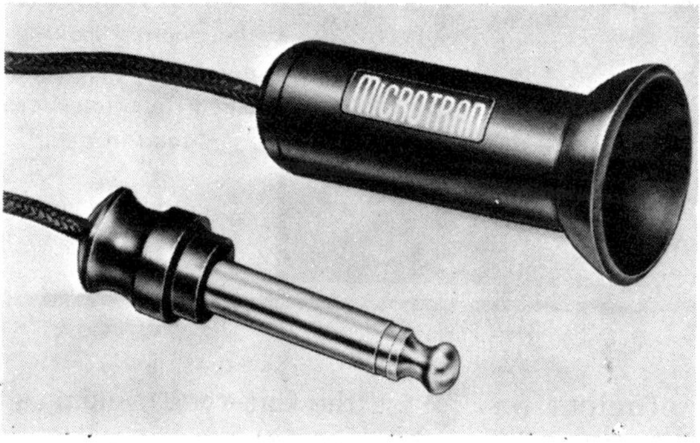

Fig. 14-1. Suction-cup pickup coil by Microtran is the most common phone pickup; it attaches to earphone area of receiver handset

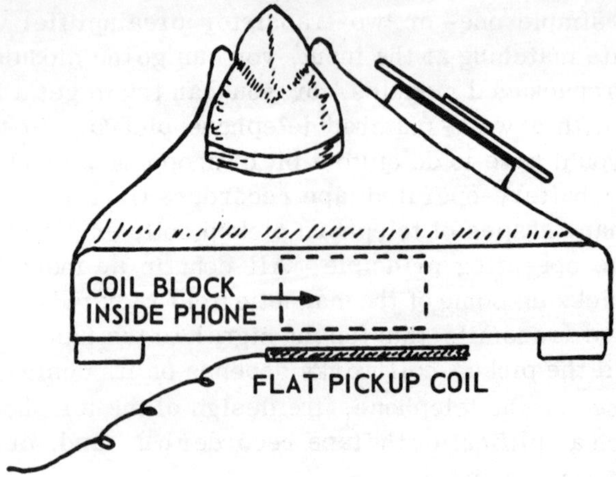

Fig. 14-2. A flat pickup coil must be placed under the phone, directly under the coil block. Move the pickup around until you find the spot with best pickup, then tape it to the phone's underside.

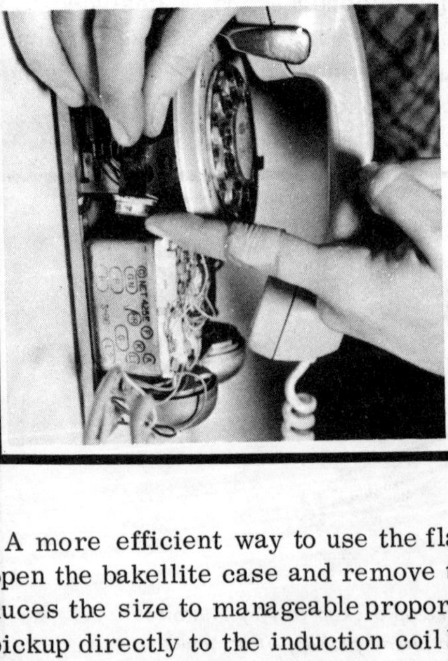

Fig. 14-3. A more effective method is to remove the coil from its plastic case and tape it directly to the block inside the phone, making a semi-permanent installation. Here, a hum-bucking Microtran coil is placed in a wall phone.

A more efficient way to use the flat-type pickup is to break open the bakellite case and remove the coil from it. This reduces the size to manageable proportions and you can tape the pickup directly to the induction coil block inside the telephone for a semi-permanent installation (see Fig. 14-3). This way there are no dangling cords or other encumbrances to get in your way when you're recording and you can tape a call at a moment's notice.

Another very useful device is the room amplifier. You can buy these already assembled (see Fig. 14-4) or make one to your own specifications from preassembled modules or other prefabricated circuits. A room amplifier offers several advantages. First of all, it lets more than one person listen to the phone conversation. Another is that it ends once and for all the very annoying need to hang on the phone waiting for someone to look up something or to get someone else. You just turn on the amplifier, put the phone down and go about your business. You'll hear the party when he comes back on the line, you can pick up the phone, and nobody will be the wiser for it. The room amplifier can be designed with a lot of extra features including a tape input so you can record when you want to.

LINE PREAMP

The basic line preamp using a $3.50 module from Carl Cordover is probably the simplest solution to recording problems. This preamp is designed mainly for citizens-band microphones, but it works like a charm with telephone recording. The module instruction sheet shows a volume control in the circuit. You

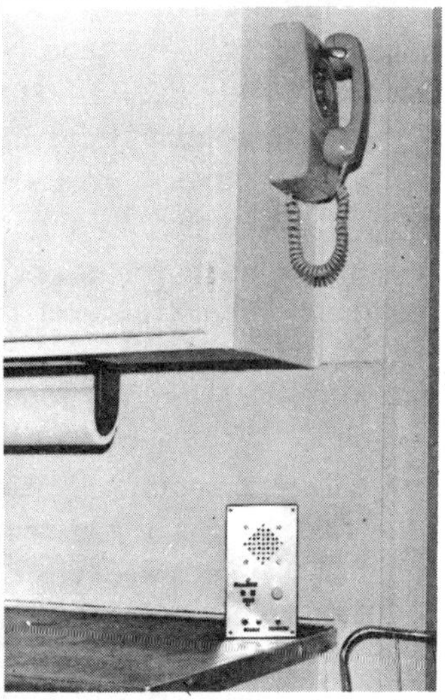

Fig. 14-4. Room amplifier attaches to the phone with a suction-cup induction pickup, frees you for hands-off phone conversations.

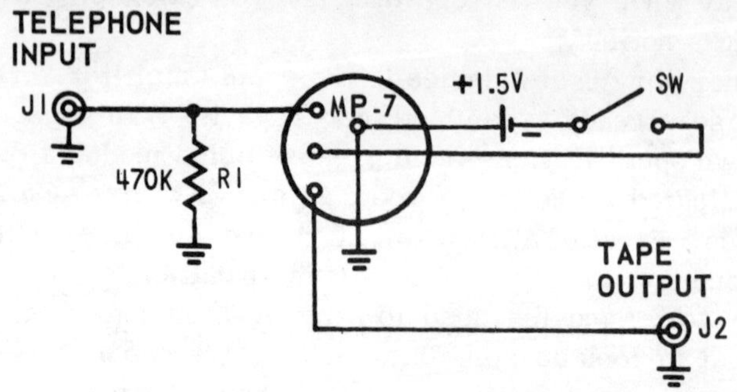

Fig. 14-5. Wiring of the Cordover MP-7 Citizens' Band mike preamp module. Jack J1 should accept the plug from the phone pickup coil. J2 is the output for the tape recorder.

don't need the volume control, since for recording you'll want all of the amplification you can get out of this gadget. Wire up with a 1 1/2-volt battery (see Fig. 14-5), install output plugs, and you're set to go.

The one shown in Fig. 14-6 uses a Minibox with two pairs of jacks—a phone jack and phono jack each at the input and output sides. The reason for this is that telephone pickup coils come in a variety of shapes and sizes and with different kinds of plugs. This setup does away with a lot of adapters and fiddling with the soldering iron later on. The only control on it is an on-off switch which also can be made to control the tape recorder if you hook it up for remote control (see Fig. 14-7).

One of the things that you'll discover with telephone recording is line hum. Hum can originate from many sources, although the most common is the AC magnetic field in the air around the telephone. This hum is present even when you make a normal telephone call, only you usually don't notice it at the time. But when you make a recording with all those extra wires acting as pickups and all that amplification, it comes barreling through loud and clear.

How to eliminate the hum? This is where a room amplifier comes in handy. With the coil on the telephone and the room amplifier turned on, wave the telephone handset around the air, trying different locations and positions to determine where hum is the worst and where it's lowest. If you're near electrical appliances and outlets, you may get lots of hum. The

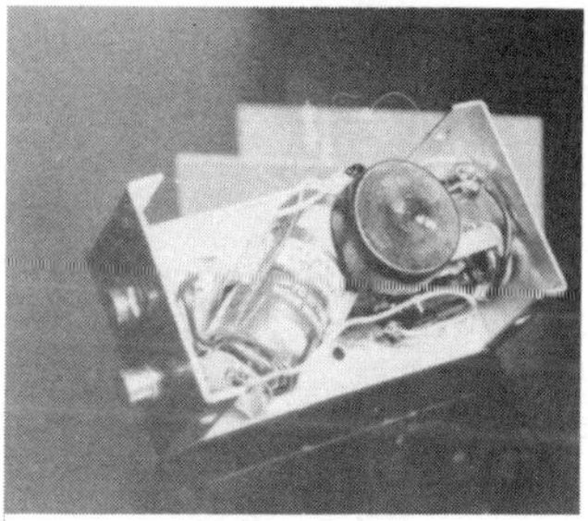

Fig. 14-6. Assembled CB preamp in a Minibox with dual jacks at both input and output sides for maximum flexibility and adapter-free operation.

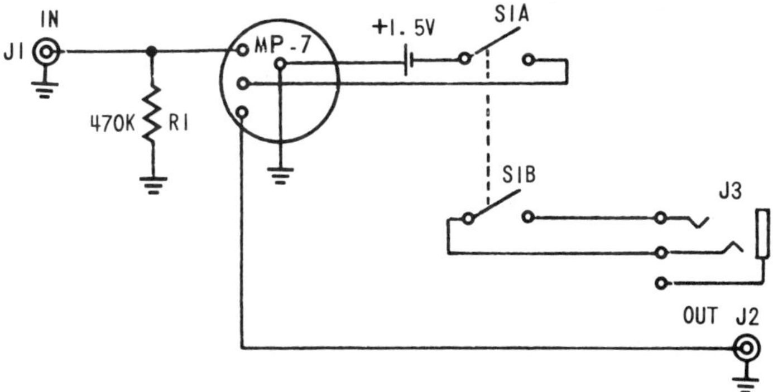

Fig. 14-7. If you use a double-pole power switch in the CB preamp Minibox, you'll also have remote control of the tape recorder.

Fig. 14-8. The hum-bucking suction-cup coil installed on this phone has two coils and two cups.

Fig. 14-9. This hobbyist is using the phone preamp with a Uher tape recorder. Battery recorders are best for phone recording, since they have the least severe hum pickup problems.

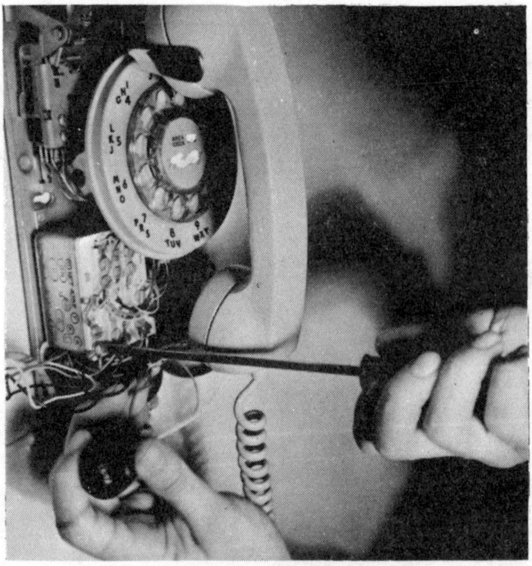

Fig. 14-10. Installing the Cordover "Conference Broadcaster" module in a phone is a simple matter. It will broadcast both sides of the phone conversation into any nearby FM radio.

moral of the story: The most convenient location for a telephone isn't always the best one for recording, so a long cord on the phone or a jack with an extension would be a big help.

Some of the hum is induced in the telephone line itself. About the only way to get rid of it is to use a hum-bucking coil, a 2-coil pickup that cancels out induced hum from one half of the line to the other (see Fig. 14-8). In practice it works very well, but the output level is a little bit lower than with the more usual type of pickup coil. Again, it's a question of how important it is to reduce this hum.

RECORDING SPEED

The best speed to use for telephone recording is 1 7/8 ips. This is slow enough to conserve tape and it's also a standard speed that can be duplicated on the big rig in the living room for later playback purposes. It's also a great convenience for high-speed tape duplicating.

For a convenient setup you'll probably want to have a tape recorder near your most frequently used phone, and you'll want to keep that tape recorder there on an almost permanent basis. If you do this, it means tieing up a hunk of equipment, so naturally you'll want to use a relatively inexpensive machine. A reel-type AC/battery recorder is usually best for this purpose (see Fig. 14-9). Use the machine on a battery while recording for minimum hum pickup. If you must use it on AC, try different positions for the machine and try reversing the line cord to get minimum hum pickup.

You may want to copy the telephone tape so you don't have to tie up those precious 5-inch reels you're using on the telephone recorder. Since you're recording at 1 7/8 ips, you can duplicate these tapes at a higher speed. In this case, high speed means 7 1/2 inches per second. Doing this is pretty simple. Put your telephone tape on one tape recorder, put your blank tape on the duplicating machine (a second tape recorder) and set them both at 7 1/2 ips. Run the tape through until you reach the end, play back the duplicate at 1 7/8 ips and lo and behold, there is your duplicated tape in exactly one-fourth the time! By the way, this is basically the same technique that's used for making pre-recorded tapes on the assembly line (see Chapter 19).

Another way to record from the telephone and provide group

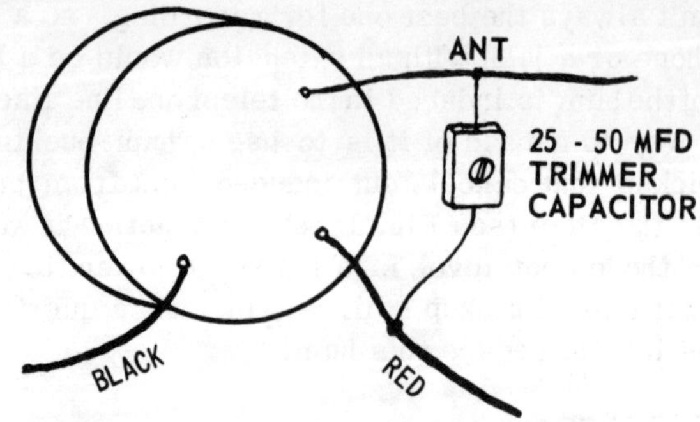

Fig. 14-11. Adding a variable trimmer capacitor to the Conference Broadcaster as shown here permits accurate frequency adjustment.

listening is an ingenious little device, also from Carl Cordover, called a "wireless conference broadcaster." (It also works equally well as a hidden telephone tap.) The $10 module is designed to operate from the telephone company's 48-volt DC power and it's very simple to install: Just remove either wire from the line to the coil block on the phone, attach the wire to one of the two screw terminals on the module, attach the other wire lead on the module to the telephone block, put the cover back on the phone, and you're in business (see Fig. 14-10).

With this module in place, you'll be able to hear both sides of the telephone conversation loud and clear on any FM radio within 200 to 300 feet. You may want to retune the module to a clear channel on the dial, either by bending the "antenna" on the module, or by adding a little trimmer capacitor to the circuit (see Fig. 14-11).

Chapter 15

Recording Film Sound Tracks

If you're a shutterbug—and most of us are in this day of idiot-proof cameras—chances are you have a collection of color slides or home movies, or both. After a few months or years of showing your silent home movies, the thought has probably occurred to you that it would be nice to add sound. Synchronized sound for home movies isn't all that new; it's just that some new techniques have evolved in the last few years that make the idea both more appealing and less troublesome.

SOUND STRIPES

A fairly common method of adding a sound track to your home movies is to sound stripe the film. This is a thin stripe of iron oxide that's added to the edge of the film after it's been processed. This sound stripe is just like magnetic recording tape; the big difference is that it uses your movie film as a base instead of an otherwise inert base of mylar or acetate (see Fig. 15-1).

Here's how it works. You get your film back from the lab and spend an evening editing and splicing it into a continuous show of say 200 to 400 feet. You might want to splice in some special-effects scenes, some footage of a color title or two. You get the thing arranged to your satisfaction, show it to the family a few times until they're all tired of it, and then put away the projector. Then you pack up your 400 feet of movies and take the reel to your photo shop. He'll send it to Kodak or another lab equipped to add the sound stripe.

At the lab, they add the magnetic coating to the film and send it back. The charge can run $15 to $25 for a 400-foot reel, but now it's all set for sound. The sound is put on through an

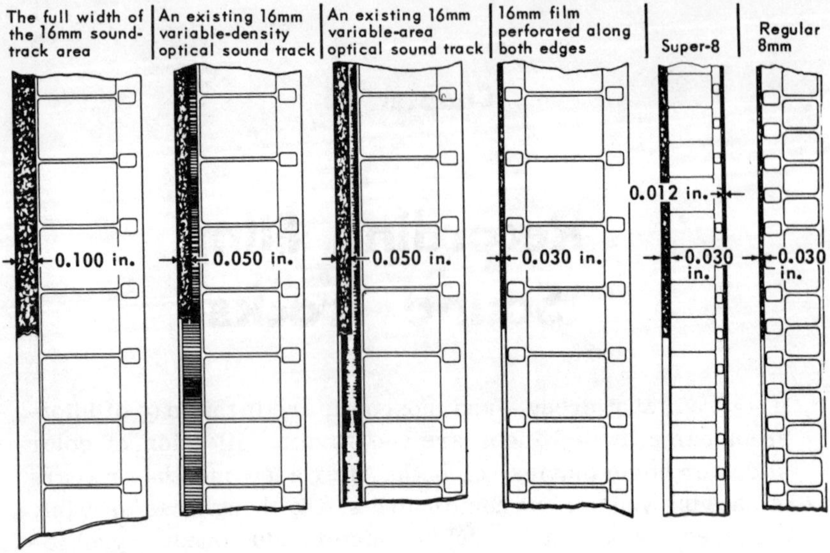

Fig. 15-1. These sketches show how Kodak sound stripes are added to various types and widths of movie film.

attachment you can add to the rear of most projectors. This attachment contains a record/playback head and erase head, along with the necessary electronics. In effect, it's the head and electronics part of a monophonic tape recorder.

DUBBING THE SOUND TRACK

You add the sound by talking into a microphone (if you're narrating some travel films or other such documentary) while the film runs through the projector. You can add background or mood music from records or tapes, piping it through a mixer. If there's dialogue in the film, try to get the people who appear in the movies to fill in the blank spots for you by talking on cue. You might want to do all this on a tape, then edit it down and dub it onto the film. Ultimately, it may be easier this way, but you have to be careful to maintain sync between the projector and the tape recorder.

Some projectors already have the magnetic sound stripe section built in, but they're usually quite expensive. Also, if you're a Super-8 devotee, you'll find that such projectors are extremely expensive, being made mainly for use in educational audio-visual work (see Fig. 15-2).

You may want to pick up local, authentic sound on the spot while you're filming. You can do this by carrying a cassette or other similarly lightweight recorder with you when you're shooting. One trouble is that if you carry the microphone yourself, it'll usually pick up the whirring sound of the movie camera shutter so well that it'll compete with the sound track material you want. Here's where an assistant comes in handy —-someone to hold the recorder and microphone a few feet away from you while you're shooting.

Once you have the tape that goes with the footage, you can edit it electronically, dubbing from the cassette onto conventional open-reel tape, leaving out the parts you don't need.

Fig. 15-2. Kodak sound projector made specifically for Super-8 sound-striped film.

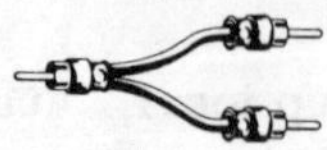

Fig. 15-3. "Y" connector used to mix stereo tape outputs into a single channel for sound-track recording.

Then you can remix, adding a sound track. A dandy way to do this is to make a master open-reel tape on a stereo recorder. Record the dialogue, getting it into sync with the movie film, on one of the stereo channels. Then go back through the film and decide what kind of music you'd like and where. Make out a script, using a stopwatch and/or the digital counter on the tape recorder for the right cue spots.

Then, when you've got just the right amount of music in the just the right places, dub from the tape onto the sound track —monophonically. Use a "Y" connector to combine the two stereo channels into a single mono channel (see Fig. 15-3). Adjust the relative volume levels at the playback recorder as you go along. In effect, your playback volume controls are mixer controls, so you can tone down the music for background under the narration or dialogue, or you can have it come crashing through full volume to add emphasis to some dramatic sequence. If you don't like the results on the film soundtrack, it's just like tape and can be erased for a retake.

TAPE SOUNDTRACK

If you don't have and don't plan to buy a sound-stripe unit for your projector, you can try using tape itself as the sound track. The process is exactly the same, except you put the final mix onto a reel of tape on another recorder, instead of on the film. The problem here is getting the film and tape to start off together and to stay together during the screening. Since the two may tend to wander away from each other during the show, it's a good idea to insert a motor speed control into the system, either to control projector or recorder speed. If your projector has a variable-speed knob on it, that's half the battle. Then you'll just have to use the control for the recorder (see Fig. 15-4). In such cases, the speed control will slow down the speed; it can't make it go faster. This method is ultimately lots cheaper than the sound stripe system, but a little more troublesome for the operator.

AUTOMATIC SYNC

A tape soundtrack system was introduced a while back by Bell & Howell. The system includes all the necessary equipment: a camera with a special synchronizing circuit; a cassette tape recorder specially modified for the sync pulses; and a projector that has a built-in synchronizing section.

The cassette recorder, slung from the photog's shoulder (see Fig. 15-5) dutifully records the sound track. It runs only when the camera runs (there's a cable connecting the two), and its second track records a series of synchronizing pulses generated by the camera. On playback, these synchronizing pulses lock onto a circuit in the projector so the tape is always perfectly synchronized with the movie film. Trouble with the system is that no one has yet answered the question of what to do with the tape if you edit and splice the film. The ultimate answer is probably another recorder—an open-reel type that can operate with the sync pulse system.

Fig. 15-4. This electronic G.E. motor speed control is helpful in synchronizing the projector and tape recorder, either during sound track transfer or separate tape soundtrack screening.

Fig. 15-5. Bell & Howell's sync sound system uses a cassette recorder with the camera. The recorder dubs synchronizing "spikes" on the second tape track for playback use.

SOUND FOR SLIDES

The idea of a sound track for your color slides may not be as far-fetched as it appears at first. Let's set the stage in your living room. You have some company over for dinner, and while you're all relaxing over coffee, out comes the slide projector. The slides may be nice, but invariably each one demands some kind of explanation. You've probably given this spiel so many times you know it by heart and are getting tired of it. Also, different slides require a longer "looking" time than others. Sometimes you want to rattle off three or four slides in rapid sequence. This rules out the automatic timer on the projector, so you've got to sit there and push that remote control button. No chance to get up and get a refill on the coffee. No chance even to be a good host—you're a slave to that pushbutton and the recitation.

This is where the sound track comes in. If you own a stereo tape recorder and have an automatic projector with a remote-control button, you can add a slide-sync system to it. Several projector manufacturers offer an accessory item priced at $15 to $25 to connect the projector and the tape recorder. Record your sound track for the slide (just as you would for movies) and when the time comes to change slides, you press a button. This puts a signal tone on the second stereo channel (the one you won't be using on playback). When the tape plays back, the tone (inaudible, since you're using just one track

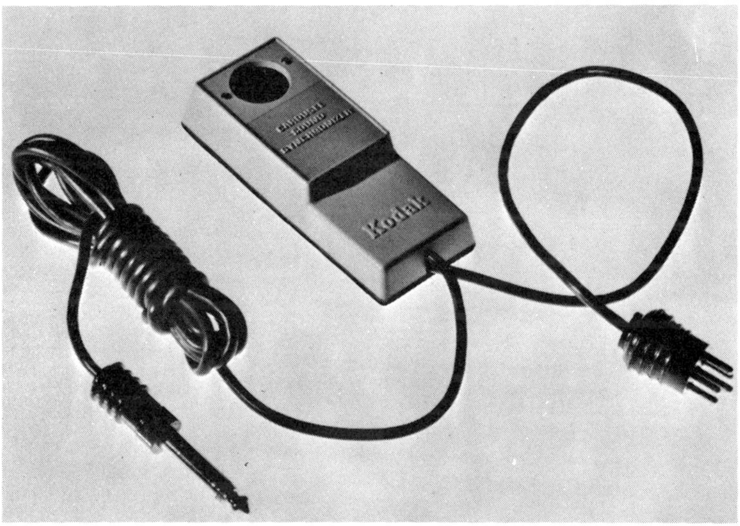

Fig. 15-6. Synchronizing control for using tape recorder with Kodak Carousel slide projector. Tape carries narrative/background sound track along with triggering pulses to change slides automatically.

for mono sound) triggers a relay which in effect pushes the button on your slide projector (see Fig. 15-6).

Now you can sit back, relax, take a snooze, get a beer from the "fridge," take the dog for a walk—nobody'll miss you and you won't get hoarse from a long-winded dissertation. The age of automation is here; you might as well take advantage of it.

Chapter 16

Maintaining Tapes & Recorders

There are certain ground rules to follow if you want to get the best performance out of your tapes and equipment. Ordinary care of the tape itself must be followed up with a small amount of regular attention to the recorder. When the tape is kept in good condition and the machine isn't, you have trouble.

Caring for recording tape is relatively simple. There are just a few basic things to keep in mind. For one, keep the tape at a relatively even temperature (ordinary room temperature is fine) and don't try to use it right after it's been subjected to temperature extremes. Tape gets awfully brittle if you've stored it in the trunk of the car on a sub-freezing day. Let the tape warm up to room temperature. The same is true of the equipment. A recorder won't work well (if at all) when it's fresh out of 20° storage. Give the equipment a couple of hours to attain room temperature before you try to use it. If you have to use it right away, carry the rig in the front seat of the car where it'll benefit from the heater.

SPLICING AND LEADERS

An easy way to keep tape in good shape is to use leader tape, a heavy-duty uncoated mylar tape (no oxide on it) that you can splice onto the beginning (leader) and end (trailer) of each reel of tape. The leader protects the recording tape against damage and breakage from repeated use, threading, rapid snap-off from the reel (during fast forward or rewind) and the other mechanical jolts and stresses the tape goes through when it starts and stops. The plain white (or other colors if you prefer) tape also provides plenty of writing surface for labeling the tape.

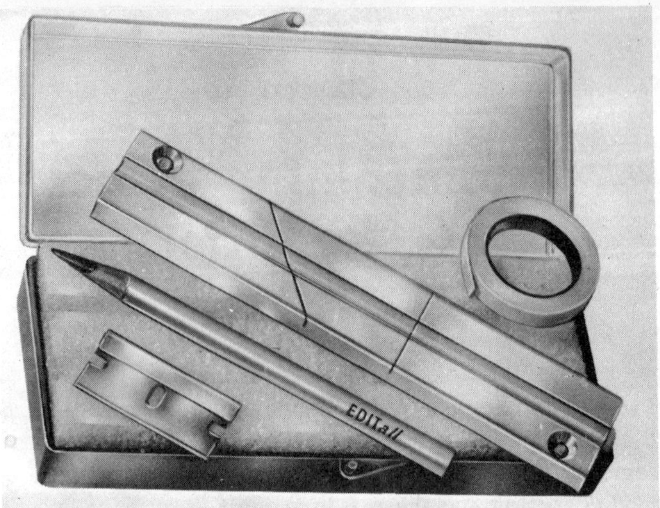

Fig. 16-1. This splicing block by Editall is the simplest and professionally preferred tool for making tape splices.

Fig. 16-2. For the all-thumbs set, a splicing machine like this Lafayette unit may be preferable. It's fast and easy to use.

Another trick is to use two different colors—say, white for the leader and yellow for the trailer. This way you instantly know which end of the tape you have at hand and whether or not the tape needs rewinding. Some manufacturers, especially 3M (the Scotch Tape people), print markers on their tape—marked off every 7 1/2 inches. This is a timing aid, so if you want exactly four seconds' worth of leader, you can measure it off very easily, even while you're threading the tape on the recorder.

Adding a leader involves exactly the same operation as making a splice. When splicing broken tape, or making an editing splice, you'll need a couple of basic tools: some special splicing tape and some kind of cutting tool. The cutting tool can simply be a pair of sharp scissors, or a razor blade. If you want to make it a little easier, you can invest a couple of dollars in a splicing block (see Fig. 16-1) or in a tape splicing machine (see Fig. 16-2).

THE SPLICING BLOCK

The splicing block has a tape-holding channel that's just slightly narrower than the tape, so the tape is held in place by friction. The block has a slot at a 45° angle—perfect for making the required 45° cut with a razor blade. That's the first step—overlapping the two tape ends and cutting _them both at the same time_ at a 45° angle. You can do this with a scissors, too (see Fig. 16-3).

The next step is to lay a piece of splicing tape over the 45° cut, with both cut ends butting flush against each other. This can be tricky without the block or machine, but it can be done (you just need five hands). The splicing tape is a special, extra-thin tape with an adhesive that will stick to the plastic side of the recording tape, but not to much else. And be sure you stick it to the _plastic (shiny) side_ of the recording tape when you splice. Burnish the splicing tape and then trim both sides of the splice, cutting into each side slightly to make a "Gibson Girl" waist type splice (see Fig. 16-3). Making the splice slightly narrower than the rest of the tape is a must, the splice will pass unnoticed over the recorder's heads and it won't get hung up in the heads or guides.

If you use a splicing machine, the roll of splicing tape is held in its own dispenser on the machine. The machine's head has two sets of built-in cutting blades—one for the 45° cut and the other to make the Gibson Girl trim. The machine also has two felt-padded arms that hold the tape securely in place—that's the other three hands you'd ordinarily need with a scissors splice!

ROUTINE MAINTENANCE

Strictly speaking, head care isn't really maintenance—it's just something you _must_ take care of at regular intervals.

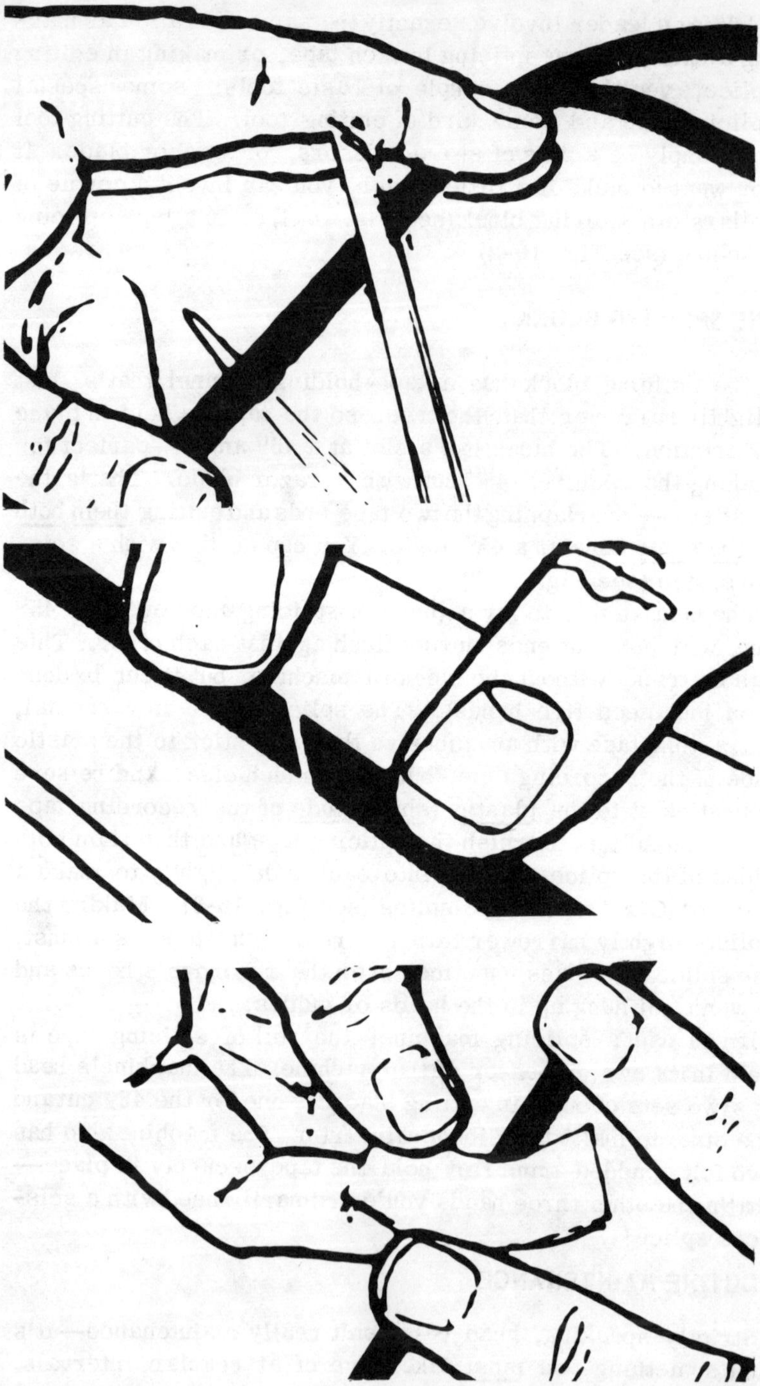

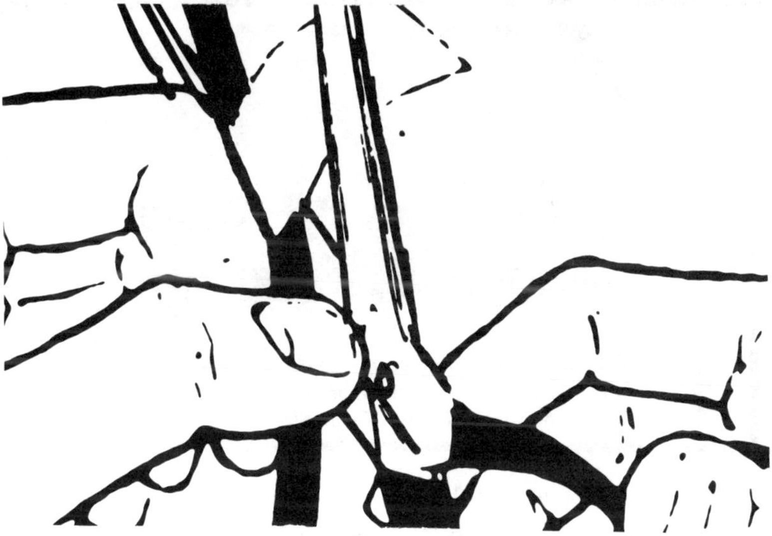

Fig. 16-3. Step-by-step scissors spike process is shown in these 3M drawings.

How often you clean and degauss the heads depends on how much you use the recorder. In above average use (say 10 to 15 hours a week), you should clean and degauss the heads once a week. Why?

For one thing, the oxide side of the tape presses physically against the heads during its passage, resulting in continuous, abrasive-like wear that causes tiny particles of oxide to flake off the tape. Yes, tape does wear, but imperceptibly so in ordinary home use. The flaked-off oxide particles catch in the head gap and tend to fill it up. Some oxide will also accumulate on the capstan, guideposts, and other surfaces that come in contact with the tape. But it's the oxide in the head gap that's the big trouble-maker. Remember, this is a magnetic material we're talking about, and clogging the head gap is like putting an iron keeper across the poles of a horseshoe magnet—the iron oxide buildup is a magnetic short circuit, drawing the head's field inward. Result is poor frequency response and increased tape noise.

CLEANING THE HEADS

To clean the heads, first remove any access or trim plate covers that hide them from view. These plates are usually

Fig. 16-4. Cleanliness is "akin" to professionalism. Keep tape heads clean with a cotton swab and carbon tet solution. Do it often!

held in place by a simple friction-fit and will pull right off. Next, using a cotton swab and carbon tetrachloride (or a prepared head-cleaning solution), swab off the head surfaces gently (see Fig. 16-4). The swab will come away with a brown coating of removed oxide particles. If there's lots of brown on the swab, let the heads dry and repeat the process with a clean swab.

The heads will dry in seconds, since carbon tet is highly volatile. It's also completely non-flammable, so there's no fire hazard. But its fumes are heady stuff and can be dangerous if you inhale very much, so keep your nose out of the work! You can buy carbon tet in some drug stores, most hardware stores, and radio parts stores. Don't get sucked in by cleaning preparations, such as Carbona, which contain carbon tetrachloride. These cleaners also contain foaming agents and other additives that aren't much good for tape recorders. Get the pure, unadulterated stuff.

DEGAUSSING THE HEADS

Another problem with tape heads is the residual magnetism that builds up with normal use. Residual magnetism is like a DC component added to the signal voltage. The net result is that the noise level gets progressively worse as residual magnetism builds up. Degaussing is an easy operation, requiring a single, inexpensive tool—a head degausser or demagnetizer. These devices sell for about four to seven dollars.

The mechanical part of the recorder—the tape transport—is a frequent source of minor problems after the machine is a couple of years old. Primary culprit here is time. The rubber parts tend to crystallize and harden with age. This means that belts, pulleys, capstan pinch rollers—virtually all rubber parts have to be checked and replaced every couple of years to maintain the machine in top operating order.

TRANSPORT TROUBLES

The degausser consists of a sealed-in coil with one or two pole pieces extending from the coil casing. An AC power cord trails out of the other end. To use it, expose the heads as you would for cleaning—in fact, it's a good idea to clean and degauss the heads at the same time. Plug in the degausser and just barely touch the head gap (ever so gently) with the pole piece (see Fig. 16-5). Then slowly withdraw the degausser from the head, moving it back from the tape recorder to a distance of three or four feet. Then pull out the plug. Just be sure there are no recorded tapes in the vicinity when you do this, since the degausser's magnetic field can damage them.

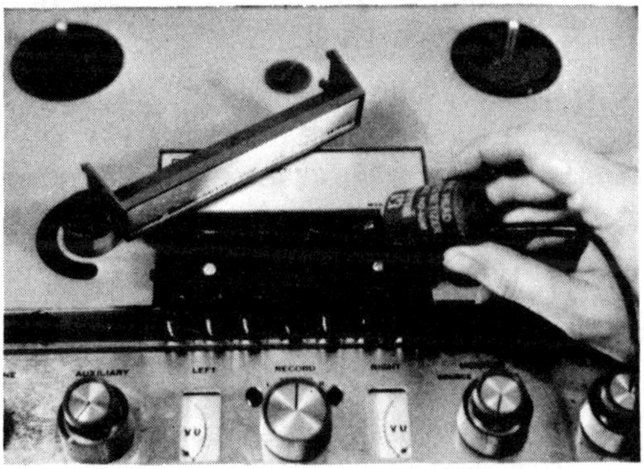

Fig. 16-5. Head degaussing is fairly simple with the proper tool—here, a demagnetizer by Micro-tran. Hold the tool's pole piece close to the head for a second, then withdraw it slowly. Also demagnetize the capstan and tape guides.

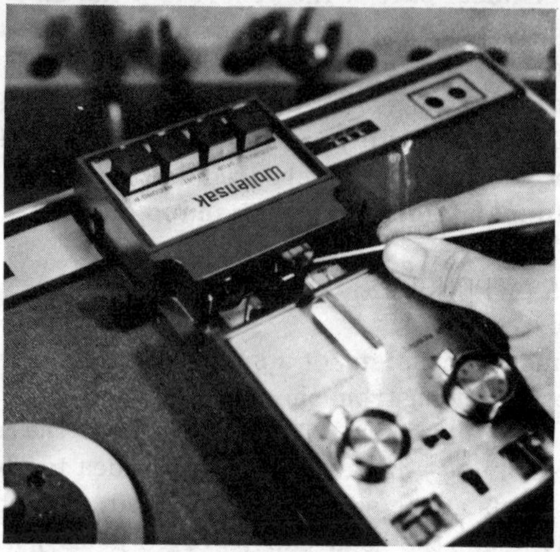

Fig. 16-6. Felt pressure pads wear and come loose, as shown by the loose pad here on the end of the stick. Replace pads before the felt gets down to the metal holder.

This problem is especially acute in home-type machines. Such machines typically have just one motor and a complex maze of levers, pulleys, belts, slides, and other moving parts "under the hood." Those myriad moving parts aren't really all that difficult to handle. With the cover off, work each control and watch how the belts and pulleys operate. If you've got a specific trouble with the machine, it's pretty easy to figure out which rubber part plays the key role.

If you're getting a lot of wow, for example, chances are the belt that snakes around the capstan pulley (or flywheel) has to be replaced. It has to stretch only a very tiny amount to cause severe wow problems. If the braking mechanism isn't working right, check the tension on those braking pads—(there are usually two for each reel). They may not be grabbing the reel pulley fast enough. A small spring tension adjustment may do the trick. Or if the pads are felt-covered, perhaps a new felt pad on each is called for.

Another place where felt pads are used on some machines is the assembly that pushes and holds the tape against the heads (see Fig. 16-6). These pressure pads wear out with surprising regularity. Renew them at least once a year, or you'll not only get degraded sound, you could damage the heads. Other machines use various methods of placing tension on the tape to hold it against the heads. Such machines have no felt pressure pads to wear out. Other belts that should be replaced at least every two years are the belts for the takeup and feed reels. Some machines may use a rubber-tired wheel to transfer power to the reel assembly. Replace those rubber tires at the same time you replace the belts.

Some belt-and-pulley assemblies are hard to get to. Very often, you'll have to disassemble part of the transport mechanism to retrieve the belt and replace it (see Fig. 16-7). Problem here is that there may be a waiting period between the time you disassemble the transport and install new belts, since your dealer may have to order the parts for you. If this is the case, keep the nuts, bolts, and miscellany in a compartmented plastic box with clear labels on the compartments so you know where the pieces came from.

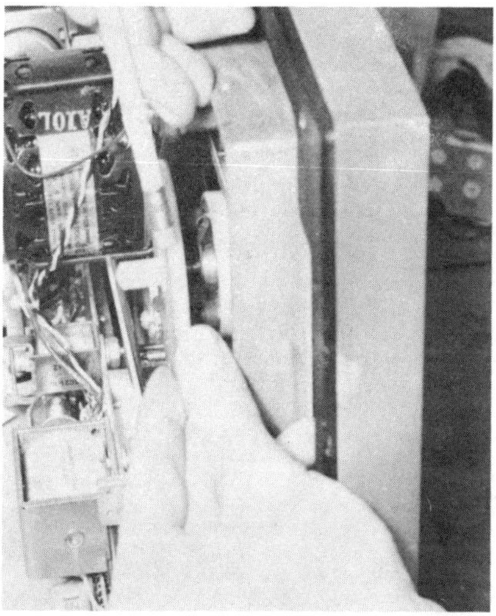

Fig. 16-7. Roughing up a rubber tire with a pencil eraser can gain a few extra weeks of service before it needs replacement.

If possible, obtain a service manual for your tape recorder before you touch it. This will show you exactly which bolts have to be loosened or removed and it will list the part numbers for the rubber. Then order a complete set of new rubber from the factory before you start dismantling the unit.

Sometimes, it's possible to do emergency, temporary repairs. If a belt slips off the capstan, it's easy enough to put it back on. Buy why did it slip off? Age and stretching most likely, and your machine will probably be showing other signs of its age. Sure, fix the belt. Maybe rub it up a little with a pencil eraser to "liven" its gripping surface. But this is only a temporary repair and its effect won't last too long. Get off that order for new rubber right away.

Chapter 17

Special Effects In Recording

An ever-present problem in recording, both at home and in the professional studio, is the relative "deadness" of a musical recording made in an acoustically dead living room or studio. Certainly, it's possible to liven up the acoustics a little bit, but you just don't get the kind of reverberation that is part and parcel of a large concert hall or theater. It is just this large-hall type of sound that is so desirable in many recordings, especially the piano and certain vocals that otherwise might sound empty or poorly recorded.

BASIC REVERBERATION UNIT

The usual way to handle the problem in a recording studio is to add a little reverberation. This is done by running the amplified sound from the microphone through a mechanical reverberation system and then to the tape recorder. The reverberation unit consists of a spring that vibrates, carrying the sound from a transmitting transducer at one end to a pickup transducer at the other end of the spring (see Fig. 17-1).

The spring in this unit carries the sound acoustically, and because there is a delay in time from the instant the sound starts at one end of the spring and reaches the other, an echo results when the delayed sound is recorded along with the original straight-through sound. The amount of reverberation that you use depends on how much you need and how you set the reverb level control. Reverberation units of this type are generally known as Hammond-type reverbs.

You may already be familiar with this reverb since it is offered as an option for many car radios today. Typically, such an automotive unit sells for about $20 to $30 and adds an ethe-

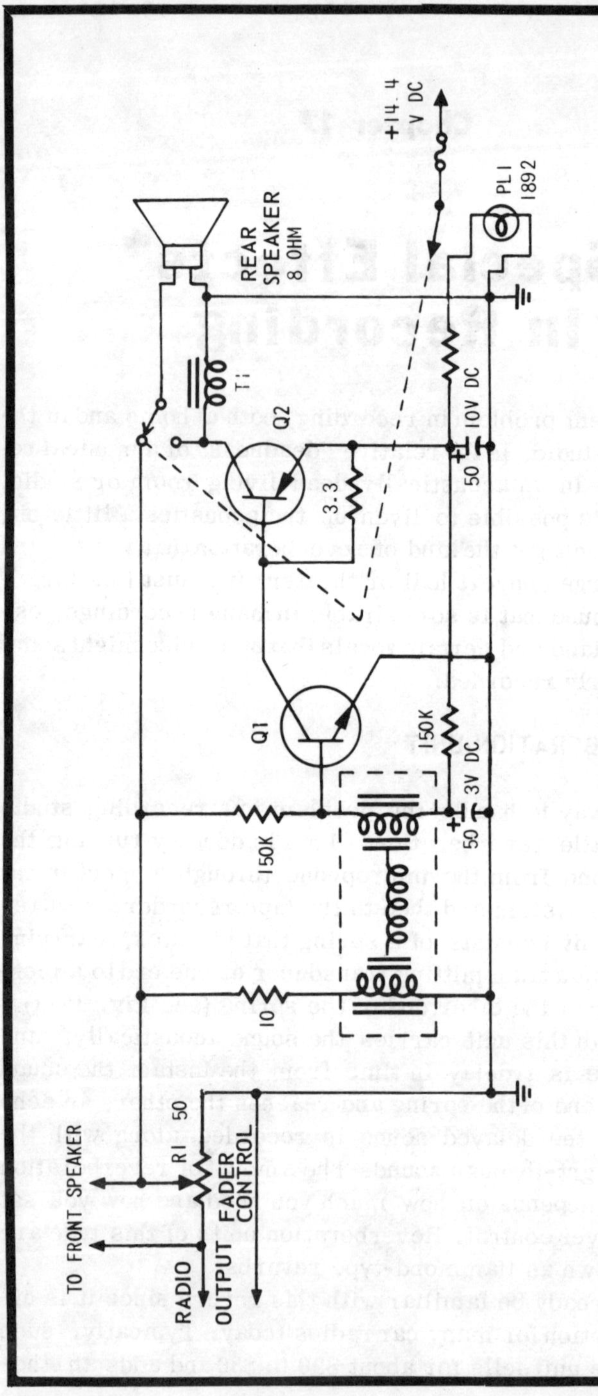

Fig. 17-1. Basic circuit for reverb spring delay line.

real, super-quality effect, even to the most lowly AM radio sound on the highway. Taken a step further, such a reverb unit can be added to your existing stereo system so that reverb can be injected artificially only on the playback of your favorite music. Once again you can adjust the amount of reverb. Units of this type are made by Fisher Radio for about $80 and other units are available at lower prices from Lafayette and Allied.

But it's in the recording itself that you may want to add reverb to get a special effect, and this is where the reverb comes into the realm of art. You can go out and buy a reverb from Lafayette for about $22. This is a monophonic unit and can be used in one channel only during a stereo recording session (see Fig. 17-2). If you want stereo reverb, then you'll need two units. This type unit is ideal for addition to your stock or stereo recording accessories.

TAPE REVERB

There's another way to obtain an echo effect with a tape recorder, but this can be done monophonically only, and only with a stereo machine that has three heads. You may recall that the third head is for separate playback and monitoring while you are recording. That third head also can be used to add reverberation to recorded sounds. If you take a single channel that's recorded on the left channel of the tape and patch the output of that channel into the right channel input, you'll re-record a delayed sound on the right channel (see Fig. 17-3). The amount of reverberation depends on what you do when you remix from one tape recorder to the other. You can play it back this way with the volume controls set differently for the amount of reverb. Or you can re-record it on another tape in the single channel, adding just the right amount of reverberation to the tape. It's a little bit complicated this way, but it can be done. One thing though, if you do this at the 7 1/2-ips speed, the echo will come much sooner after the original sound than it will at the 3 3/4-ips speed. If you want a long, drawn-out echo, then use the slower speed.

There are commercial reverberation units on the market that use just such a tape delay system. These reverb systems have six or eight heads or more to provide a selectable amount of delay. These machines have a continuous tape that is used

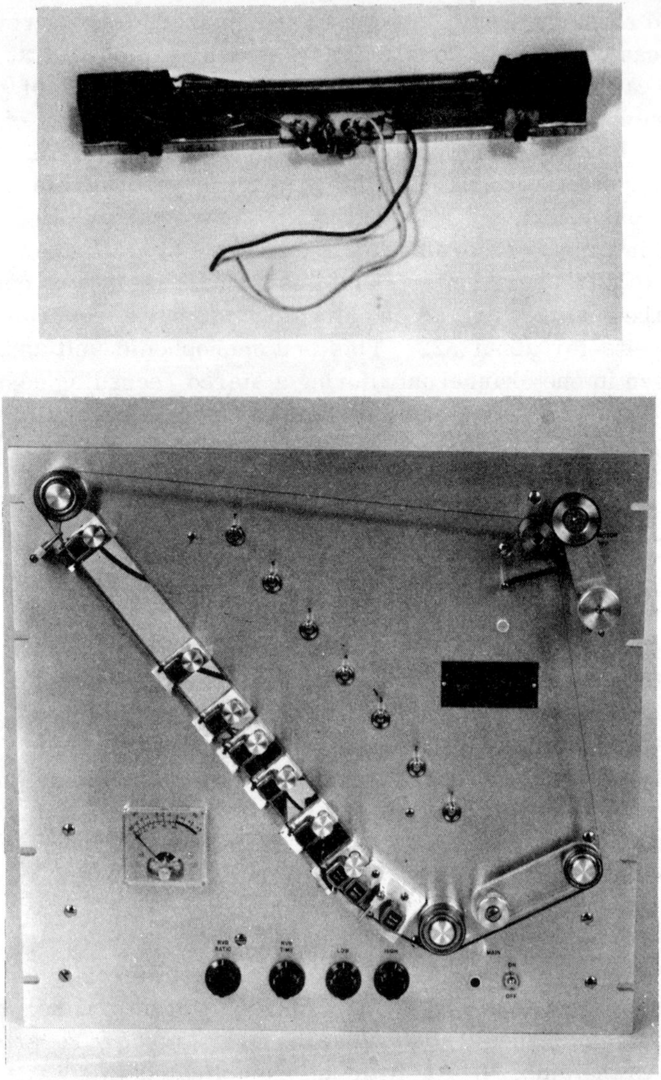

Fig. 17-2. (A) Prepackaged Lafayette reverb unit for home use—either during playback or making recordings. (B) Tape reverb unit uses endless loop of tape passing over series of adjustable heads for any desired amount of reverb time. Tape is erased and reused continuously in this device made by Audio Instrument Co.

over and over—the reverb is added and used and then everything wiped clean for the next pass around.

If you want to build a reverb system, you can pick up the mechanical equipment—the Hammond springs—from Olson Radio for a couple of dollars each. For five dollars, you'll have the spring equipment to build a stereo reverb accessory. The circuitry involved with these accessories is pretty straightforward.

CENTER MIKE FILL

Another type of effect that you may want to experiment with is something that the professionals call a "panpot." This is a control that can move a microphone's apparent position in the stereo playback. You've no doubt heard recordings where an instrument moves electronically from the left through the center and to the right of an orchestra and back again. While this zooming, panning microphone appears very cute, the most frequent use for a panpot is to place a microphone electronically in a precise position for best stereo balance.

If you want to work with three microphones, you may not get exactly the kind of balance you want for the third mike using an ordinary mixer. One way to work this is to bridge the microphone output through a pair of audio transformers and then feed each transformer output into a mixer input (see Fig. 17-4). If you use a 4-input mixer such as the Switchcraft model described in Chapter 11, you can get an interesting mix and movement effect, but here you're limited to three microphones, one for each stereo channel and one for the center or "panfill."

You may not even need the matching transformers, depending on the characteristics of the microphone that you use. You could conceivably get away with a "Y" connector which plugs it into the left and right inputs simultaneously.

PANPOT CONTROL

Another way to do it: Simply hook up the microphone to a 500,000 ohm potentiometer (linear taper) and bridge it to the two stereo channels. This requires two Y connectors or mounting in a Minibox with input and output jacks. The Minibox method is probably just as cheap and easy to assemble as are a pair of Y connectors, and the results will probably be more reliable and satisfying.

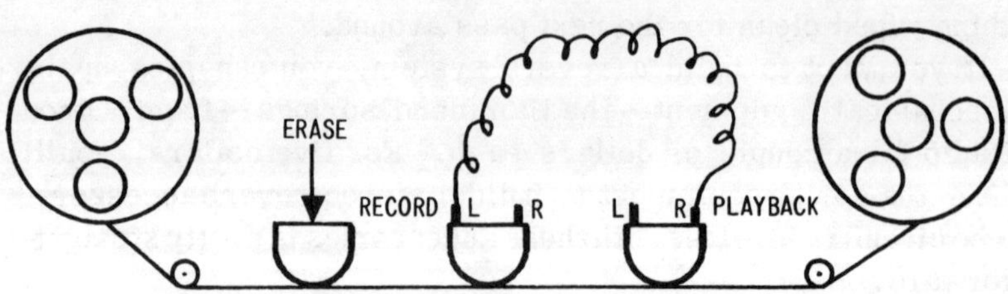

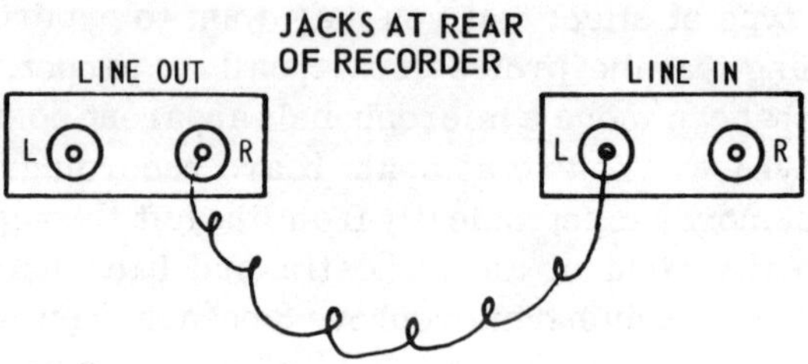

Fig. 17-3. Wiring a three-head tape recorder as shown here will add reverb to a recorded tape. This setup works for mono recordings only on a 4-track stereo machine.

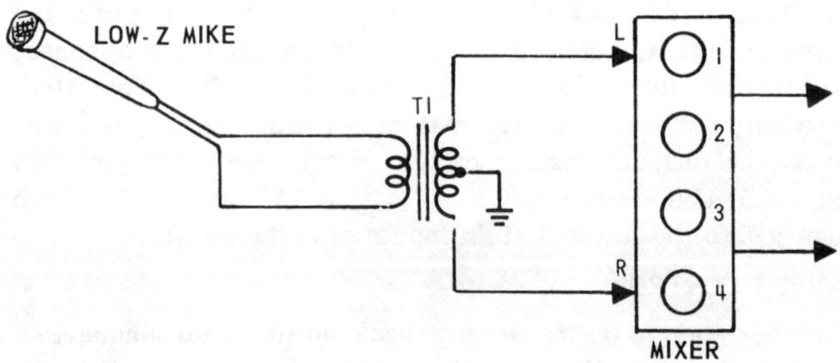

Fig. 17-4. One quick way of adding a center microphone with a simple mike mixer (stereo) and on audio transformer with a high-impedance, center-tapped secondary.

The whole point is that you just can't add a third microphone for center fill as easily as you can add a third loudspeaker to your playback system. It'll take some juggling, it'll take some careful watching of those impedances, and a little experimentation, but you'll be more than satisfied with the results.

One thing you should try to avoid is a matched microphone for the center channel. Try to select a microphone that has a higher signal output than the two matched mikes that you use for the left and right stereo channels. This means either knowing your microphone characteristics very thoroughly or going out and buying a microphone specifically for the center-channel fill. Failing that, you'll have to add a preamp with volume control to the panpot control box (see Fig. 17-5).

TAPE EDITING

You've probably heard about tape editing. It's very much like movie film editing, since the tape is a ribbon-like medium. All that's really needed for editing is the basic kit of splicing tools (see Chapter 16). Editing usually involves removing un-

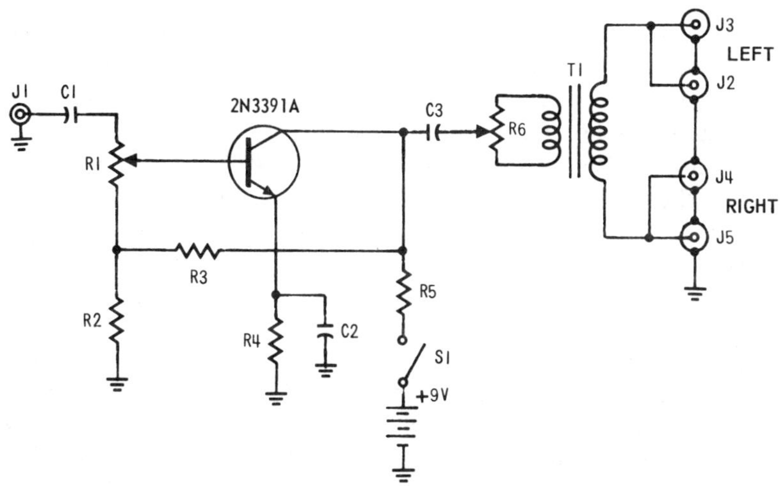

Fig. 17-5. An amplified panpot control provides the extra gain needed for the center fill mike, with full adjustment range for volume and apparent mike position.

wanted material from the tape. It can also mean moving something from one section to another. It can mean putting together bits and pieces until you have constructed a conversation or statement by someone—something he never really said. If you have enough taped conversation material from a person, you can diddle with the razor blade and splicing tape until you virtually have the guy saying anything you want.

This gambit is used fairly often in the spy thrillers and "cops-and-robbers" programs on TV. But it's precisely because this can be done that recordings of any kind are no longer admissible as evidence in a court of law. After all, once you've made all the cuts and splices it's pretty easy to dub the whole thing onto another reel of tape. You can swear this is the original and there's no way you can prove it or that the court can disprove it. Stalemate.

But an edited tape can be very useful for a number of reasons. Suppose you have a party in your house and you leave the tape recorder turned on all evening (unobtrusively, of course). You've got three or four microphones "sprinkled" around the place and you get a fairly complete tape of everything that went on—all the fun highlights, and all the long, dull stretches.

You listen to the tape the next day and would like to save it, but without all those long, dull stretches. You just want to save the fun part. What to do? Easy. Edit it. This is where another kind of editing can come in handy—its called electronic editing, and the tools are simply two tape recorders. To edit electronically, just duplicate from one tape recorder to the other (see Fig. 17-6). Let's call the playback machine recorder A and the recording (or duplicating) machine recorder B. Set them up just as though you were making a tape duplicate, which in a way you are. Now turn on recorder A and play through the dull spots until you reach a fun spot. Stop A, rewind it maybe two or three feet, just before the beginning of the fun spot you want. Setting up the tape to a spot just before the very beginning of the material you want is called cueing. It's a common practice in radio stations and recording studios.

Now set recorder B in the "record" mode and start it. Next, start recorder A. As soon as the fun spot ends, shut off recorder B. And go on playing recorder A until you reach the next spot you want. When you've finished, you'll have an edited duplicate tape—one with just the party highlights and none of the dull, in-between stuff. At this point, you can erase the original, overlong tape if you want to.

One big disadvantage of electronic editing: Like any tape duplication process, the dubbing adds to the tape noise level —about 3 db of noise for each generation or stage of duplication. What's a generation? The original is the first generation. A dub from the original is the second generation. A dub from the second is the third generation, and so on. When you buy a record or a prerecorded tape, it's about a fourth or fifth generation because of the special processing that's needed before the final version can be released. That's why the background noise level can get pretty fierce on some prerecorded tapes.

There's a disadvantage or two to cut-and-splice editing, too. For one thing, you can't edit a tape that's been recorded in both directions. If you try editing side A of a 4-track stereo tape, you'll destroy the sequence on side B; when you cut the tape, you cut through two different programs—the one you're editing and the one on the other side. So the raw to-be-edited tape can be two-track stereo, full-track mono, or four-track stereo—recorded one side only.

Another disadvantage: All those splices tend to weaken the tape physically, and you might have some trouble if it's to be replayed very often. Here again, you'll have to dub the edited tape to get a copy that'll stand up to any amount of wear and tear. Also, once you've butchered that original tape, you won't want to erase and re-use it, since it's bound to be very

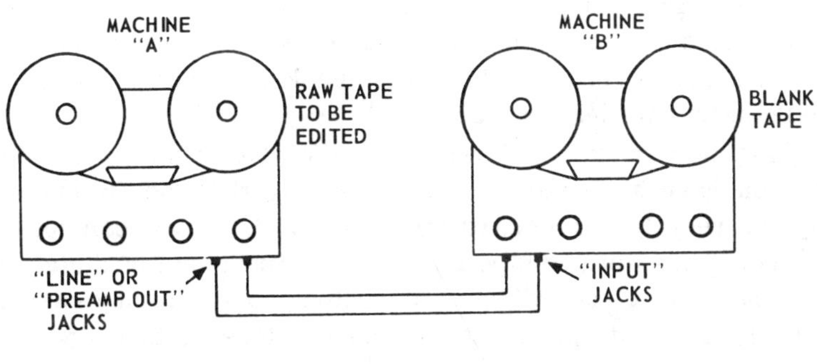

Fig. 17-6. Setup for electronic tape editing.

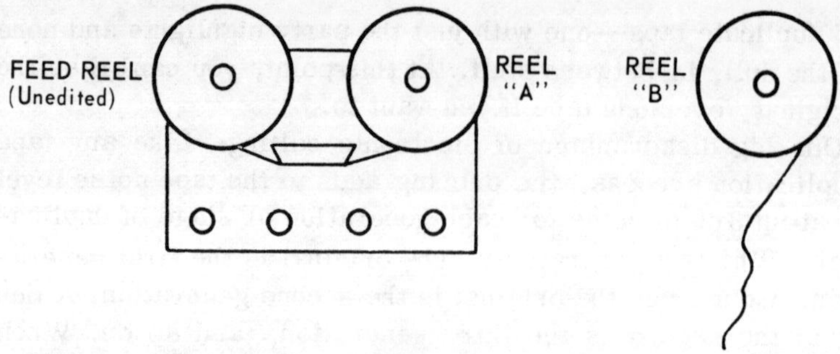

Fig. 17-7. An extra reel can be used to store edited sections of tape.

unreliable for mastering with all those splices on it. So there's an economic factor, too.

FINDING THE SPOT

If cut you must, arm yourself with a grease pencil (or Crayola crayon) and a recorder that permits easy access to the heads. On playback, stop the machine instantly when you hit the syllable or spot you want to splice. Hand-wind the tape back, listening for that exact spot. Then, mark the spot with the grease pencil, being careful not to get any of the wax on the head.

Next, pull the tape off the machine—leaving the feed and takeup reels intact. Pull out enough tape to reach the splicing block and make your cut. Presumably, the section that follows on the feed reel will be removed. To do this, remove the takeup reel with its portion of tape up to the cut. Call this reel "A." Put on an empty takeup reel and start the second section of tape on here (see Fig. 17-7). Call this reel "B." Wind through until you reach the spot where you want to make your next cut. Mark the spot, make the cut and remove reel B from the recorder. Put reel A back on, make the splice, and continue to the next spot.

This time when you make the cut and replace reel A with reel B, you have a choice: You can just lay the tape around the tape on reel B and discard this tape later, or you can splice it and use the spliced tape later as a utility tape for dictation or other short recordings. We belong to the string-saver school, so we'd splice the unused portions of tape and try to use them later for something else.

Chapter 18

Foreign Recorders & The Future

Tape recorders aren't just the shiny gadgets that you see in radio parts stores or the hi-fi showroom. They come in many forms and sizes. There are the enormous video tape machines described in Chapter 20. There are very tiny pocket-size portable recorders. And many different forms in between.

One of the smallest and most convenient recent entries is the Philips Mini cassette recorder (see Fig. 18-1). It uses basically the same type of cassette described in Chapter 2, except that it's much smaller. The cassette is not very much larger than a book of matches and holds 10 minutes of recording time on each side. The recorder is slightly larger than a pack of cigarettes and it is great for taking notes on the go, dictation, and other applications where a tape recorder in the pocket would come in handy. This machine uses just a standard 9-volt transistor battery, so you're not weighed down with a heavy battery supply.

THE NEWELL TRANSPORT

The newest tape transport on the scene today is the Newell tape system. This is an entirely new concept in tape recording and more and more companies are beginning to use the Newell transport one way or another. Right now, it's used mainly in instrumentation recording, but the day may not be far off when the newell will be in every home both for audio and for TV tape recording.

Most of today's tape recorders put an incredible amount of mechanical strain on the plastic strip that is the recording medium. Consider what happens: The tape goes past tape

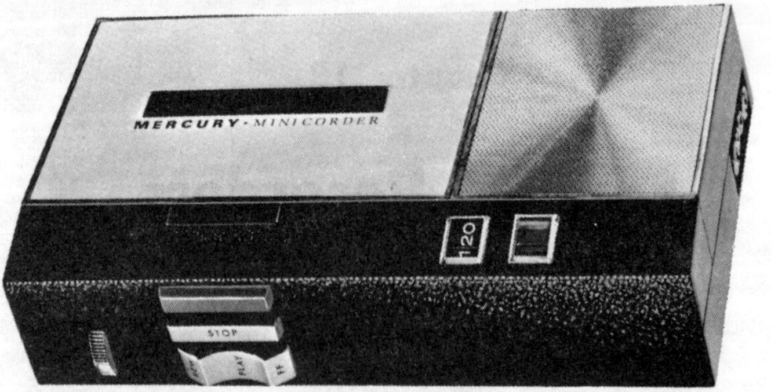

Fig. 18-1. Hardly larger than the cassette itself, this Mercury TR1000 recorder fits into the pocket for on-the-go dictation and note-taking. Other models by Philips use a smaller cassette of recent design.

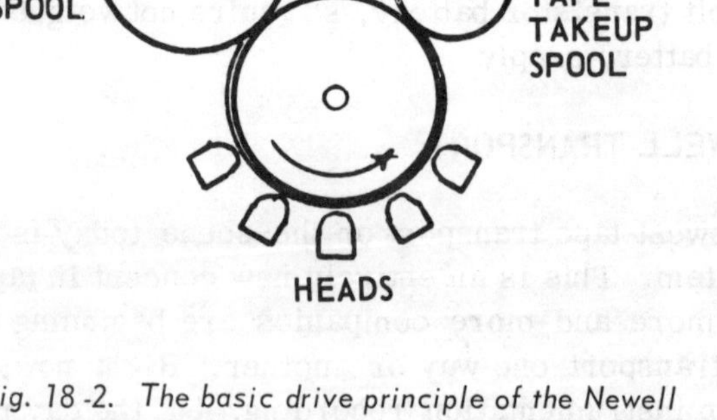

Fig. 18-2. The basic drive principle of the Newell tape transport removes all mechanical stress and tension from the tape. Tape reels or "reelettes" move on spring-loaded slides to maintain continuous intimate contact with the oversize rubber capstan.

heads where it is pressed hard into place, it is then pulled through the capstan and pinch roller, over a series of guideposts and tension bars, then onto a takeup reel that places further strain on the tape, since it always tries to turn faster than the tape is feeding through the capstan assembly.

In ordinary use, a tape can break, stretch, or otherwise be damaged in the process of recording and playback. This problem of mechanical strain becomes even more acute in systems that demand very high tape speeds, such as tape duplicating facilities (which are described in Chapter 19) and in the proposed linear-scan television tape recorders that may someday find their way into your living room. (The problems of television tape recordings are detailed in Chapter 20.)

The Newell system is quite simple in its basic concept. The reel of tape is adjacent to a very large oversized rubber capstan (see Fig. 18-2). As the tape goes around this capstan with no pressure roller against it, it passes the tape heads which are arranged around the outside circumference of the rubber roller. It then goes directly onto the takeup hub which

Fig. 18-3. One of the first commercially-available recorders to use the Newell transport concept, this "Alpha" recorder by Sangamo Electric Co. is made for instrumentation use.

is also right next to the capstan. There are no tape guides, no pressure pads, no pinch rollers, no takeup reels with their added tension to stretch the tape or cause any problems. Both the feed and takeup hubs are on spring-mounted slides so that as the diameter of the feed reel decreases, the center of the reel moves down the slide, always keeping the tape right next to the capstan. The reverse action occurs with the takeup hub. As its diameter increases with more tape wound on it, the takeup hub moves farther away, always with the tape making direct contact from the capstan to the outside layer of tape on the reel.

HIGH-SPEED TAPES

One of the great advantages of this system is that the tape can move at very high speeds with no mechanical stress placed on the tape. Another big advantage is that it can stop, reverse, and get up to full speed in the reverse direction all in just a fraction of a second. The result is that the amount of information that can be packed on the tape is enormous (see Fig. 18-3).

An ordinary 7-inch reel of tape could hold perhaps 16 tracks of television information. The first track would be played at high speed, then the tape would reverse and play the second track, then reverse and play the third track and so. Thus, a full-hour television program could be put on a single 7-inch reel. In audio recording, a very tiny reel of tape would hold an entire album or set of albums. The reel, or "reelette" as the Newell people call it, would be about the diameter of a half-dollar, and would be very convenient for storage and lugging around.

Chapter 19

How Tapes Are Duplicated

There's a plant outside Chicago that in many ways is a pioneering production facility. It's given over largely to high-speed duplication of prerecorded tapes for the consumer market. It's a big business these days, and workers at this Ampex plant operate on round-the-clock shifts because of the enormous demand for their product. In this plant are made conventional open-reel tapes, cassette tapes, and both 4- and 8-track tape cartridges. Ampex is a multi-label operation—duplicating tapes for about 50-odd different record companies and labels—individual companies who don't have enough tape business to make it worthwhile to set up their own facilities.

THE MASTER TAPE

The first step in the process is the original recording from which a master tape is made. A record company, such as London, DGG, Vanguard, Mercury, etc., sends a one-inch wide master tape to Ampex after the release or production start up on the disc version of the album. When it reaches the plant, it's reprocessed for the duplicating equipment; special cueing signals are added, both at the beginning and end; special packaging, album covers, program notes and tape reel labels are printed, and production begins.

In the open-reel production line, a master tape will drive 10 slave duplicators at a time (see Fig. 19-1). This means that each time the master is played, 10 prerecorded tapes are made on the assembly line. Blank tape is threaded from an enormous "pancake" of blank tape onto the plastic reel with the label already pasted in place. All four tracks are recorded simultaneously.

Fig. 19-1. Master tape machine feeds ten slave recorders in this typical Ampex production line.

The "A" sequence, or first side, is recorded backwards and the entire operation is done at about four times the normal speed. The master's normal speed is 15 inches per second, so this runs at 120 ips. The open-reel tapes, if they are to be 7 1/2 ips speed, will run at 60 ips. Thus, an album with a playing time of 40 minutes can be recorded in five minutes.

This would seem to be a lot of time for a girl operator on the line to have nothing to do. Actually, many of the tapes are done so quickly that she really keeps hopping. To start the line, she pushes a button, and tapes all up and down the line start whizzing past their recording heads. As soon as the 5-minute period is up, the master tape stops automatically and all of the other machines down the line stop, also automatically. The operator then pushes a button which starts the master tape on its rewind. While the master is rewinding she runs down the line, and with a deft motion of her fingers breaks off the tape on each recorded reel and tosses the reel onto a shelf above the duplicating machine. Next, a new empty reel is slipped on the machine, and with incredible speed she threads the end of the blank tape pancake onto it, then moves onto the next position. An operator can complete the entire line of 10 machines in about a minute or less if she's especially good.

In the meantime, the master tape has automatically stopped rewinding and is all set for the next run. The gal returns to the control position, pushes a button and starts the line again. Some of these operations are so fast that one gal per line is not quite enough. So they'll use three women for every two lines of 10 machines. The ones who do long classical tapes have a little more time on their hands and will busy themselves with monitoring and testing in other areas.

CARTRIDGES AND CASSETTES

Tapes for cartridges and cassettes are duplicated in much the same way, except that the recorded tape is wound onto another huge pancake continuously with a cueing tone recorded at the end of each album. These large pancakes of recorded tape then go to another part of the assembly plant where they're loaded onto a machine that senses only the cueing signal. It stops there, the operator snips off the tape and threads it onto the hub or cartridge assembly (see Fig. 19-2). It then goes to another station where other components are added to the cartridge or cassette, leaders spliced onto the tape, metal cueing foil added for cartridges, and the outer shell of the plastic case attached (see Fig. 19-3). Another machine heat-seals the plastic cases.

After this, the tapes are packed into cardboard containers (or plastic containers for the cassettes) and are then wrapped in plastic sheeting which is then heat-shrunk onto the case (see Fig. 19-4). This is the final step in processing the tapes for commercial sale. A large tape duplicator like Ampex will maintain a warehouse stock of tapes of all labels for drop shipping directly to dealers. This type of operation lets small record companies market prerecorded tapes at virtually no cost, only a royalty payback arrangement.

THE HISS-KILLER

The old, familiar gremlin of tape hiss can creep into the professional master just as easily as it can find its way into

Fig. 19-2. A plastic hub is placed on each end of a cassette album tape as it is cut from prerecorded pancake. (Courtesy: Ampex)

your living room tape recorder. The pros are concerned about a couple of other problems too—rumble and scrape added by the recorder, and print-through—especially in tapes to be stored for long periods.

The problem-solver is a "black box"—an electronics package with no knobs or adjustments (see Fig. 19-5)—made by Dr. Ray Dolby in his tiny shop in England. His black box is called the Model A301 noise reduction system. It does more than just reduce noise; it pushes it back to the vanishing point.

Thus, a new word has been added to the audiophile vocabu-

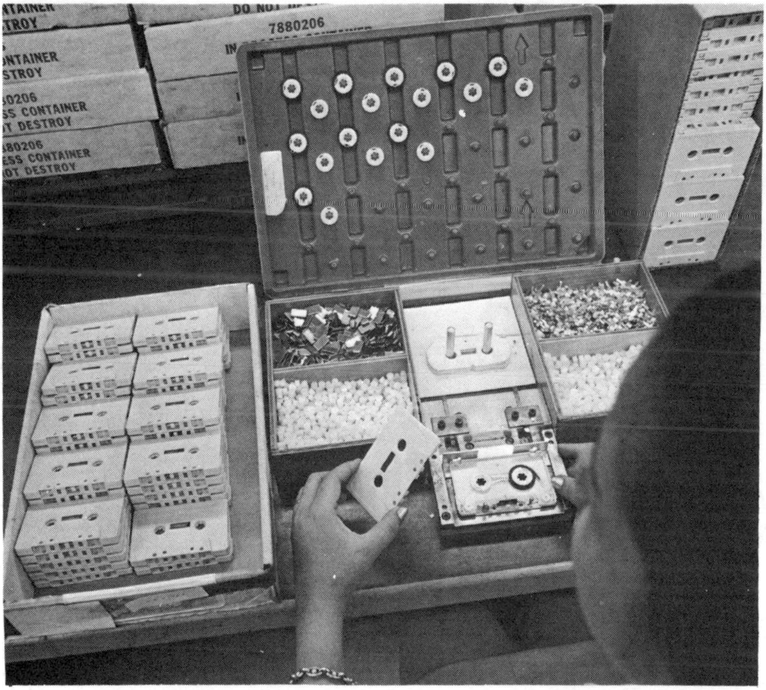

Fig. 19-3. Cartridges and cassettes are all assembled by hand. Here, hubs containing prerecorded music are placed in containers. (Courtesy: Ampex)

Fig 19-4. Final step is to cover each tape package with transparent plastic film, making the package dust and moisture-proof for storage and shipping. (Courtesy: Ampex)

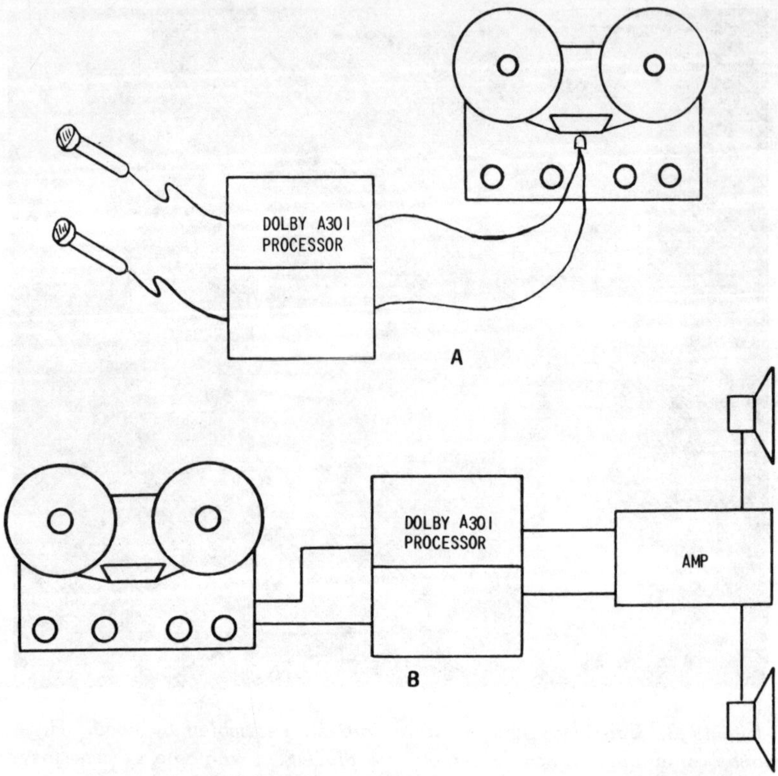

Fig. 19-5. Commercial tapes are made quieter by using the Dolby A301 processor in all steps of mastering. Any tape-recorder produced hiss, rumble, scrape, print-through, and other associated mechanical noises are eliminated (actually cut down about 10-14 db) by this technique. Final tape is not "Dolbyized" but has benefits of vastly reduced noise and hiss.

lary—"dolbyize"—which means to process tapes with the Dolby system. Based on the premise that noise is added to the tape <u>at the tape recorder</u>, the black box processes the signal on its way from the microphone to the tape. On playback, the tape must be piped back through the same black box (or any A301).

HOW IT WORKS

Noise levels—especially tape hiss—become most noticeable at low signal levels. When the music is low and soft, that hiss comes barreling through. When the music is loud, it tends to mask the hiss and whatever other noise that might be on the tape.

The Dolby processor boosts the relative signal level of very low passages while leaving the loud passages of music relatively untouched (see Fig. 19-6). Then, when noise creeps into the tape, the music is already at such a high signal level, the noise won't affect it.

In playback, the low passages are pushed back down to their normal levels. When this happens, the noise is pushed entirely out of the picture. Noise reduction is typically 10 to 14 db, and you're benefiting from it right now with some of those new records and prerecorded tapes mastered with the Dolby system. The final, undolbyized product sounds crisp and clear; all the dub-added noise is gone. The only noise present is in the final recording.

The Dolby system cures some other ills, too. Among them are print-through and crosstalk (interference between adjacent tracks on the tape.) The elimination of print-through—a major long-term storage problem—is a boon for record companies and broadcasters who store tapes for long periods of time.

There are other types of noise-reduction systems in use, with varying degrees of practicability. Very often, the compression function in such systems causes a characteristic "swishing" sound in the playback. The attack (response) time of such systems is often a problem, and the total noise limiting isn't as effective as it should be.

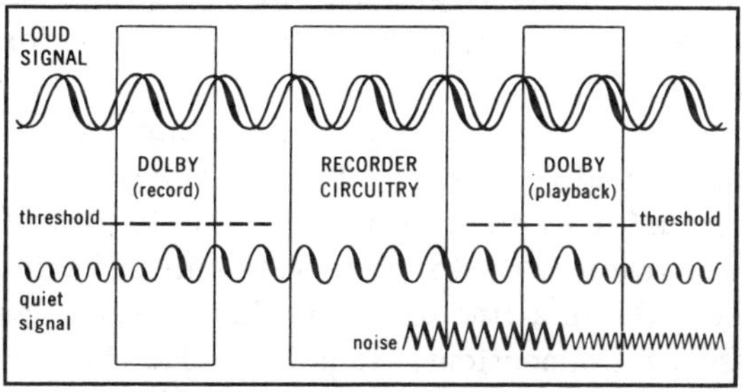

Fig. 19-6. These waveforms illustrate how the Dolby works in the KLH recorder. Low-level music is expanded or pre-emphasized to bring it well above the hiss level. On playback, the low passages where hiss is usually apparent, are completely hiss-free.

While the basic Dolby is too sophisticated for any but the recording companies, the home tape enthusiast can have Dolby benefits in his own system. One tape recorder—KLH—has one band of Dolby noise reduction built in. The machine is priced at $600, a high price for a home unit, but with that single-band Dolby circuit, it's quite a bargain. The KLH recorder's noise-reduction circuit operates only in the high band—from 2,000 to 20,000 Hertz. That's where the hiss is, and KLH Dolbyized tapes are remarkably free from hiss. To the serious recordist, this machine can be a boon.

MAKING THE MASTER

The professional studio puts a tape through several processes before a record is finally made. The raw tape from the various "takes" may have eight or sixteen channels recorded on it. Somehow, these many tracks must be boiled down to just two for commercial stereo sales. The multi-track original goes to a mixing room where an expert recording engineer tries different combinations of channels until he has just the right balance for a final tape. This "mixdown" process is done with the vast majority of today's commercial tapes, especially for pop and rock.

In the mixdown, the engineer can add special touches: echo, mechanical reverb, feedback, vibrator, fuzz, rhythm, sound effects, and anything else that comes to mind. It's an art, this mixdown. It's out of the audio engineer's monitor loudspeakers that the music first takes form and comes alive. This process has given the recording engineer a new art form, which he uses to the hilt.

This new master tape, prepared electronically, is duplicated and the dub is held in storage. The master goes to the cutting room for transfer to disc. Here, the tape is played on a professional recorder that's wired into the cutting head on the disc lathe. The lathe is a fine precision instrument, made either in Germany or Connecticut. It is here that the LP album (or pop single) sees its first light of day. The master disc cut here goes on to spawn several generations of "mother" molds and thousands (or perhaps millions) of descendents. But it is, after all, the tape that holds the key—the tape that first records, that can hold 16 channels, that can be edited to produce all sorts of special electronic effects.

Chapter 20

Video Tape Recording

Television tape recording has been with us since Ampex first introduced it to the broadcasting industry in 1956. In one swoop, the bad-quality kinescope TV recording was pushed out of the picture and "live" TV shows were dutifully recorded on tape before being put on the air.

A TV tape recorder is an all-electronic system that puts out a video signal every bit as clean as a live camera signal. It employs the same basic technology as the audio tape recorder, but with lots of extras thrown in. For one thing, to record a TV picture requires a tape system with very wide frequency bandwidth—about four million Hertz (4 MegaHertz). Remember, we learned in Chapter 1 that a home recorder has a bandwidth of about 15 to 20 kiloHertz, and for a higher frequency response we'd have to increase the tape speed.

BANDWIDTH NEEDED FOR VIDEO

A video tape recorder, to achieve a 4-MegaHertz response, should have a tape-to-head speed of a thousand inches per second or more. Speed like this is impractical, so instead, the head is mounted on a wheel which spins at a very high speed as the tape goes past it (see Fig. 20-1). The spinning disc contains four heads, so this system is referred to by broadcasters as a quadrature-head or simply a quad machine. The track recorded on the tape is very nearly perpendicular to the tape, so the scan is known as a transverse scan.

The scanning lines aren't perfectly vertical because, while the head assembly is spinning, the tape is also in motion, so the tracks are slightly slanted. Notice in Fig. 20-1 that as

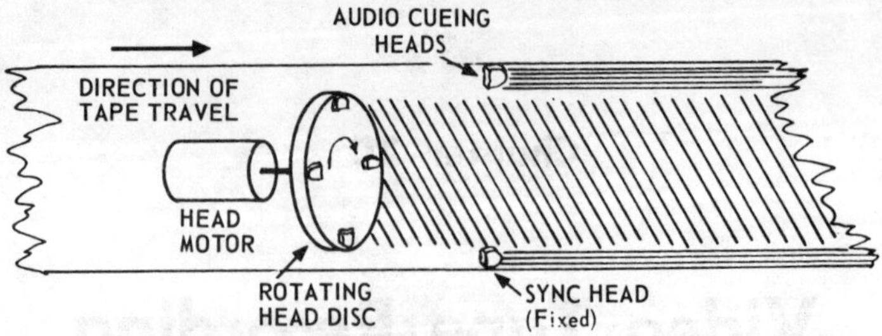

Fig. 20-1. Quadrature (four-head) video tape recorder used by broadcasters has four heads in a rotating disc which spins against tape at high speed. This raises effective tape-to-head speed to thousands of inches per second, providing more than 4 MegaHertz bandwidth. Audio, sync, and cueing signals are recorded down at tape edges by stationary heads.

the head finishes its scan, another head has already started its track across the tape. The tape speed is usually about 12.5 inches per second, and combined with the spinning heads and some very sophisticated electronic circuitry, the machine can record and reproduce excellent color television pictures.

IN-LINE TRACKS

The spinning heads don't use the entire width of the special 2-inch wide video tape. Along the edges are special tracks, recorded by conventional stationary heads, that carry the audio, cueing signals, and synchronization signals. Equipment of this kind, designed to be used in a TV station, is very expensive. A basic video tape recorder (see Fig. 20-2) is priced at $50,000 to $60,000, plus a few thousand more for needed extras. There are just two American companies making such equipment: Ampex and RCA, the latter under license from Ampex.

HELICAL SCAN MACHINES

While video tape was an instant success among broadcasters, there was a vast untapped market for low-cost equipment. In 1960, Japanese manufacturer Toshiba (Tokyo Shibaura Electric Co.) introduced a more economical system known as helical scan. This system also uses rotating heads, but the heads rotate horizontally (on a vertical axis) in the same plane as the direction of tape travel (see Fig. 20-3). This scanning technique, also known as slant-track, depends on the tape having a diagonal wrap around the spinning cylinder that holds one or two tape heads (see Fig. 20-3). Since the heads are spinning in a horizontal plane, this diagonal path is what slants the track. If you look at a helical machine (see Fig. 20-4), you'll see that the takeup and feed reels are not on the same level with each other; the takeup reel is below the feed reel.

Fig. 20-2. Ampex VR-2000 video tape recorder, workhorse of the broadcast industry, costs from $50,000 to $80,000, depending on options.

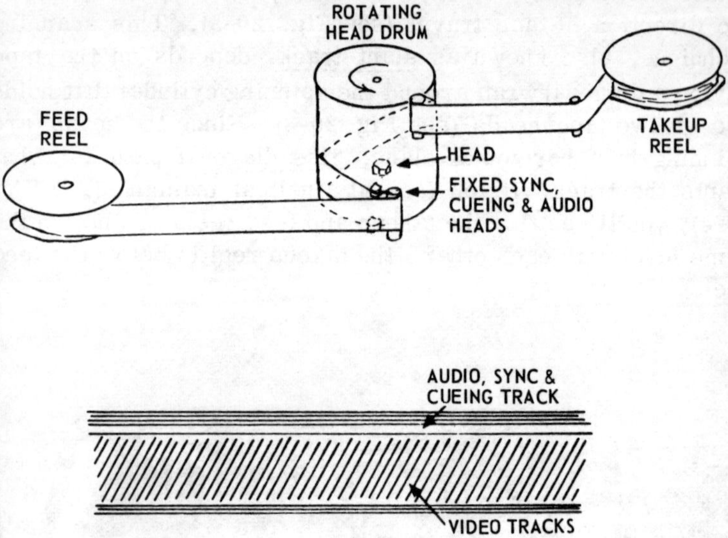

Fig. 20-3. Helical recorders use a rotating drum with one or two heads pressed against a tape that slants diagonally around the drum. Stationary heads record audio, sync, and cueing signals at the tape edges.

The first helical machines were priced at \$10,000 to \$12,000, and used one-inch wide tape, just half the width of broadcast video tape. Like the quad machine, helical recorders lay down parallel tracks along the edges with fixed heads for sound, sync and cueing. Today, helical recorders can be purchased at prices varying from \$12,000 down to about \$600, depending on how much picture quality and versatility are required. Frequency bandwidth is nowhere as high as with the quadrature machine, but it's possible with some of the better-designed helicals to get excellent color.

OMEGA WRAP

One of the problems with most helical recorders is the fact that they can't record the entire TV picture. The tape can't get an entire pass by the head since the tape doesn't completely encircle the head assembly (see Fig. 20-5). Depending on just how much of the scan is missing from the tape, there is a loss of from 6 to 12 lines in the recorded picture. This "white space" is usually placed at the bottom of the TV picture so it'll be hidden by the TV set's cabinet and screen mask. If you misadjust the TV set's vertical hold and let the picture roll, the missing lines will show up instantly (see Fig. 20-6). Some recorders, notably the International Video machine, use an Alpha wrap (see Fig. 20-7) so no scanning lines are lost.

MISSING INTERLACE

Television pictures use a two-in-one presentation known as interlace. What this means is that the first time through a single frame, only every other line is scanned. Then the in-between lines are scanned to make up the complete picture. Many of today's home TV sets don't even provide the interlace or second field of in-between lines. Others may provide it when the set is new, but the interlace disappears before the

Fig. 20-4 Helical recorders by Ampex, color model VR-7800 (left) costs $4000-$5000, while monochrome model VR-5000 (right) is tagged at about $900.

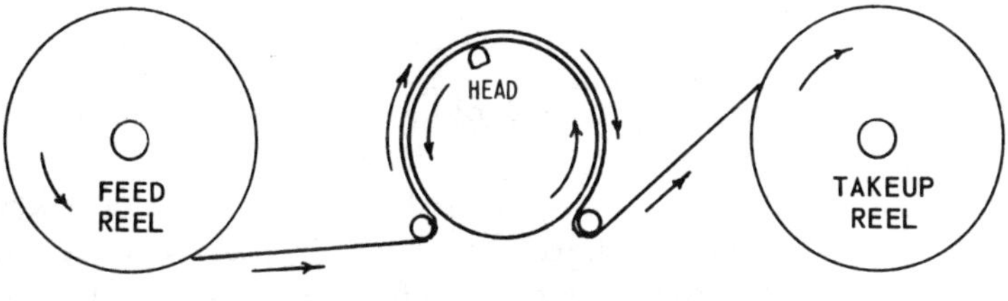

Fig. 20-5. The "Omega" wrap typical of most helical recorders does not wrap the tape completely around the drum assembly, which results in some lost TV scanning lines.

Fig. 20-6. Lost scan lines are placed at the bottom of the picture for minimum inconvenience.

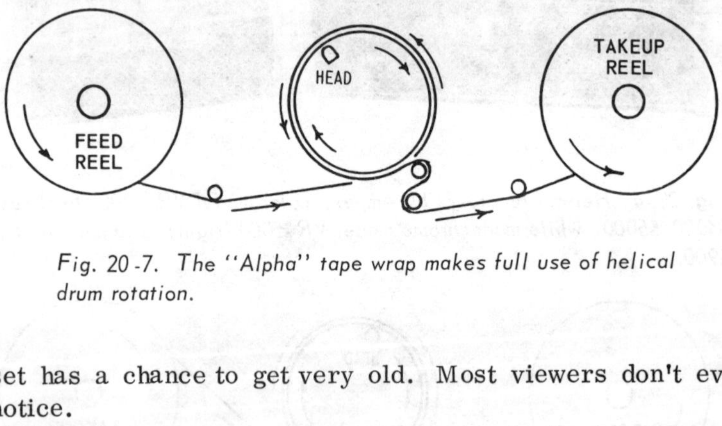

Fig. 20-7. The "Alpha" tape wrap makes full use of helical drum rotation.

set has a chance to get very old. Most viewers don't even notice.

When a small-screen set is used, this lack of interlace is even less noticeable. Therefore, some low-price VTR (video tape recorder) manufacturers capitalize on this fact, since the lack of interlace capability automatically reduces the size, complexity, and price of the equipment. It's fine as long as a small-size TV screen is used, and some VTR companies include a small-screen TV set with the VTR for just this reason. Another reason is even more compelling—the TV set must have special circuits for recording and playback; the VTR signal has to come from somewhere.

Many VTRs have several outputs, including video and RF. The RF output is a modulated signal on some commonly used VHF frequency, such as Channels 5 and 6. This option is offered, so you can pick whichever channel isn't used in your area. Then you can hook up the VTR output to your TV set's antenna terminals through a 75-ohm/300-ohm matching transformer and look at the VTR playback on your favorite TV set. Using an ordinary TV set as the recorder's signal source usually requires some modifications to the TV.

LINEAR SCAN

Aimed mainly at the potential home market, the linear scan VTR is anything but a positive technical success. This type of scan uses fixed heads (the same as a conventional stereo recorder) with the tape whizzing past at a breathtaking 90 to 120 ips. This is about as fast as the tape moves during fast forward or rewind on the usual stereo recorder. The bandwidth resulting even at this speed isn't too exciting—about 2 Megahertz—but it's enough to provide at least recognizable black-and-white TV pictures (see Fig. 20-8). Recorders of this type generally use 1/4-inch tape for economy. Several companies have announced or are developing such systems. But so far, there's nothing on the horizon that looks like a good bet for the average homeowner.

VIDEO TAPE IN THE HOME

The best bet for a "home" VTR right now is the low-end helical machine. This format is at least within reach of the luxury market and is getting better and cheaper every day. The VTR can change your whole household routine if you let it. It's a great gimmick for recording those favorite TV shows that you'd otherwise miss for one reason or another.

There's been some talk about setting up the VTR with a timer to turn it on when your favorite show comes on. This way, you'd have a recording of the program and could watch it when you get home. But it won't work—at least not now—since the VTR requires several operations to set it up and get it "perco-

Fig. 20-8. Linear-scan TV recorder by Mastercraft uses stationary heads and 1/4-inch audio tape whizzing by the heads at 90 ips speed.

lating." It's just not something that can be started with the flick of a single switch. But the idea is a basically sound one. You can still use the VTR for TV shows that come on when you're too busy to watch, or are watching another favorite show that conflicts with it. Then, instead of watching the late movie that starts at 11:30 and keeps you awake until 1:30 with a grade-Z flick, you can watch a favorite TV show during that hour after the 11-o'clock news sign off.

Trouble is, this late-night delayed viewing gets to be habit-forming. You don't want to tie up a $50 roll of videotape with that show, so you watch the program that same night, come what may, to free the tape for the next recording. After a few nights of this, you'll stagger into work with bloodshot and baggy eyes, and some of your co-workers may tactfully suggest a local chapter of Alcoholics Anonymous. At any rate, you'll get lots of fun and pleasure out of the VTR—especially at parties, when you can unlimber the camera and microphone and make some electronic home movies with instant soundtracks. The playback always gets to the crowd!

SERVICE PROBLEMS

Then one day, the rig goes on the fritz. It looks like something simple—something you, with your vast reserve of knowledge gained from this book, can fix easily. So you get out your VTVM, your soldering pencil, long-nose pliers and dikes and an assortment of screwdrivers, and looking very professional (to your wife) you proceed to make the VTRs problems ten times as bad. Don't laugh—it can happen! So then, you do what you should have done in the first place—call the service agency for that particular machine. After several weeks of waiting for delivery of your repaired machine, you'll be presented with a bill. One word of advice: be sitting down when you open the bill to take a peek. It'll be a lulu.

Nobody's trying to rob you. Even the least expensive video tape recorders are incredibly complex machanisms that require technicians of an extremely high caliber. There aren't enough of these techs to go around, which you may have already discovered if your TV set has ever gone into the shop. The shortage is nowhere more acute than it is in the VTR service industry.

Yes, the VTR is lots of fun, but be prepared to shell out again and again: first the purchase price; then for expensive video tape; then for the very necessary servicing which can come fairly often. Your alternative might be to wait a couple of years when linear-scan systems will be available using the Newell transport. These are in the works now and it may not be very long at all.

Chapter 21

The Snoopers Around Us

On a busy street in our nation's capital, not too far from the White House, is a tiny store that unabashedly proclaims its role in life. It's called the "Spy Shop" and in its window is a variety of electronic eavesdropping devices ranging from the well-known Martini olive to the most sophisticated types of electronic listening devices known.

But a device used for electronic eavesdropping also may have many less-sinister applications. Certainly the wireless FM microphone would not be considered a spying device. Yet, all it takes is the simple act of hiding an FM microphone under a table, and it has suddenly become a "bug". The main difference between most wireless devices and those used for eavesdropping is the fact that the latter are hidden.

TAPE RECORDER IS THE KEY

A key element in any eavesdropping operation is the tape recorder. Without the tape recorder and its voice-actuated circuit, the professional (and amateur) eavesdropper would be relegated to the role of Elliott Ness' cohorts—round-the-clock listening to a lot of nothing on a pair of uncomfortable earphones just to catch one all-important telephone call. But the wonders of modern science have made all of this unnecessary. The modern, technically up-to-date snooper makes his rounds, changing batteries and reels of tape, covering as many as a couple of dozen listening posts in the space of a couple of hours. Once he's changed the batteries and the tape, he knows he'll get at least a solid day's continuous monitoring with absolutely flawless recording of whatever transpires.

To "bug" a room, the snooper first must have access to it.

Once inside, either with a set of passkeys or with a few well-placed $5.00 bills, our snooper finds the most likely places to secrete wireless mikes. Ideally, there should be a few windows in view, since the building's steel skeleton will absorb a great deal of the transmitted energy of these tiny snoopers, proventing him from getting a good pickup on his FM receiver. With his bugs in place and turned on, and all telltale signs of his entrance covered up, the snooper quietly leaves, snapping the latch behind him. Suddenly, a very safe apartment or office has become a very dangerous place to hold any sensitive conversations.

What about batteries? Yes, they do run out, even on these expensive little bugs. But chances are, the snooper will have all the information he needs within a day or so—well within the lifetime of the batteries in his equipment. In many cases, the snooper, detective, agent, or whatever else you want to call him, may have to write off the cost of the listening devices, since he may not have another opportunity to enter the premises to retrieve them.

What happens on the outside? Somewhere within a few hundred feet of the stakeout, our hero sets up a battery/portable FM radio that's been modified for the lower frequency band used by the clandestine transmitters. The frequencies are just below the standard FM broadcast band and any standard FM equipment can be altered very easily for these frequencies simply by adding a small trimmer capacitor in parallel with

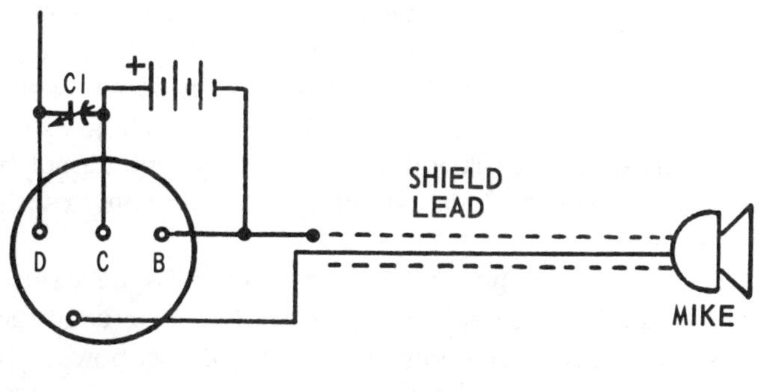

Fig. 21-1. A Cordover wireless mike module can be tuned by placing a 25-mfd trimmer capacitor across the "antenna" and positive battery leads.

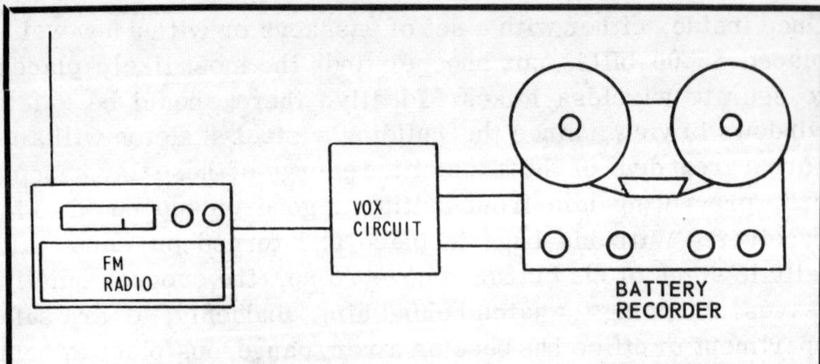

Fig. 21-2. Receiving end of an automatic bugging setup—a battery FM radio and battery tape recorder with a VOX circuit.

the tuning circuit on the transmitter and the receiver (see Fig. 21-1).

VOX CIRCUITS

The radio is turned on and it will play for a few days on its battery supply. Connected to the radio is a battery-operated tape recorder with a VOX circuit (see Fig. 21-2). Turn the tape recorder on, activate the VOX, and the recorder will run only when someone is speaking. Result: great savings in tape and in time that ordinarily would be wasted listening to blank tape.

Supposing it's the telephone that our bugging expert is after. Again, he must gain access to the premises. Once inside, he can place any number of gadgets inside the telephone. The simplest ones contain their own batteries and are planted inside the telephone case (see Fig. 21-3). These will transmit a signal, the same as the wireless bugs, including both sides of the telephone conversation. Bugs of this type do not break the telephone circuit, but instead pick up the conversation by inductance.

A fairly common type of telephone tap does necessitate a physical break in the telephone circuit. It's inserted in the line and runs on the telephone company's 48-volt DC power. Since this transmitter has no battery to run down, it'll operate indefinitely, broadcasting its signal to a nearby FM receiver (see Fig. 21-4). This type of bug has varying degrees of acceptance. It can be detected because of its drain on power.

But it doesn't need constant attention to keep batteries in shape.

LONG-DISTANCE BUG

A relatively new entry in the telephone-tapping field is the "Infinity" sold by several electronic eavesdropping supermarket operations. This device comes in two parts. The main section is installed in the telephone, as with any tap. The other part is a coding beeper which is carried by the eavesdropper. Here's how it works. Our hero has planted a bug in a phone in Los Angeles. He's in New York and would like to know what's going on in that room in LA. He goes to a private phone, dials the area code and the number, listens carefully for the various circuit relays to click into place, and just after the last relay is connected and before the first ringing signal occurs, he presses a button on his beeper. This releases a shrill note on a precise frequency. The reed relay in the companion unit at the other end of the line responds instantly, cutting out the bell-ringing circuit in the telephone

Fig. 21-3. This self-contained phone tap is small and difficult to detect, and has its own battery supply.

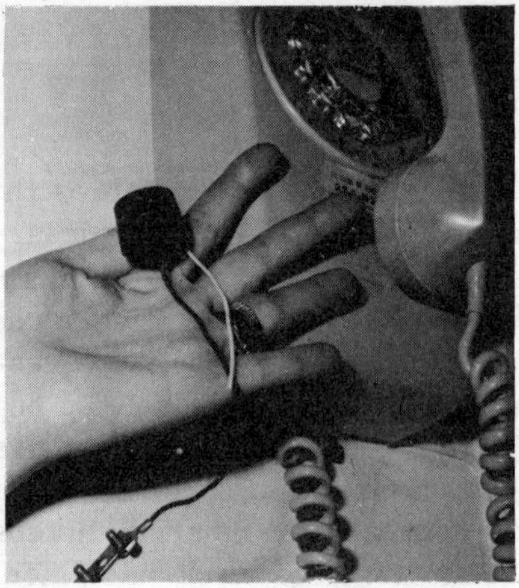

Fig. 21-4. Wireless conference broadcaster module plays both sides of phone conversation into an FM radio up to 200 feet away. This $10 module's frequency can be changed with a trimmer capacitor and it becomes a telephone tap that's permanently powered by the phone company's power.

and turning on its own self-contained microphone. Our eavesdropper can then listen to a conversation in the bugged room, simply by listening to the telephone.

When he's through, he hangs up. Since the call wasn't completed, there's no charge from the phone company. We're not certain how well the system works, although we have it on good authority that it does indeed do what it's supposed to.

RUSSIAN BUGS

There are other, more sophisticated eavesdropping devices. For example, there is the famous Great Seal bug that the Russians placed in the U.S. Embassy in Moscow. It was a wireless microphone that transmitted nothing! The microphone consisted of a flexible metal diaphragm connected to a waveguide cut to a specific microwave resonant frequency (see Fig. 21-5). When the diaphragm vibrated, the resonant cavity length changed slightly in time with the vibrations. The Russians established a listening post across the street from the embassy. They pointed a high-intensity narrow-beam microwave signal at the Great Seal, received, and recorded the reflected wave that they got back from it. This reflected wave contained all the conversations carried on within earshot, or should we say within microphone-shot. The only way the bug was ultimately found was by dismantling the entire embassy, piece by piece.

THE "ANTI-BUG"

For every bug, there must of necessity be some kind of "anti-bug." To detect the presence of a telephone tap there are various checks that can be made, and there are devices that can be played through a telephone to jam sensitive conversations. The trouble is they jam a lot of what you want the right person to hear also.

A widely accepted solution to the bugged telephone syndrome is the scrambler. A scrambler is a device that electronically mixes up the telephone conversation into an unitelligible jumble. At the other end, the second party has another scrambler with the same electronic code as the one that you're using. This unscrambles the garble, while at the same time garbles the conversation coming in the other direction.

Scramblers may be purchased in large quantities—dozens or more with exactly the same electrical code—if this is necessary. At least one major European automobile manufacturer issues scramblers to all its key executives to prevent industrial spying. Apparently, they had a bad experience a couple of years ago with overheard telephone conversations and have been using the scramblers ever since.

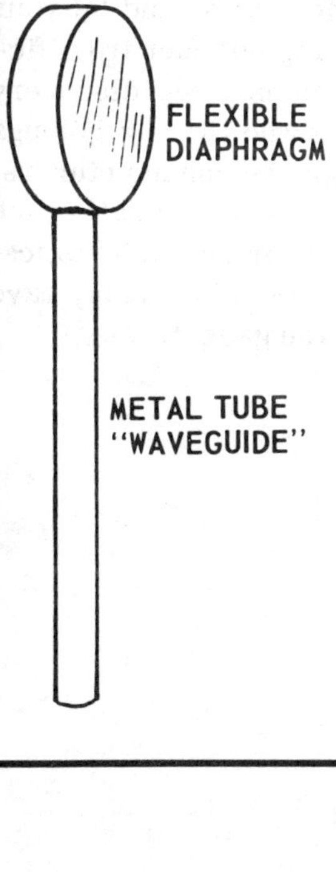

Fig. 21-5 A passive microphone of the type sketched here was used successfully by the Soviets to bug the U.S. embassy in Moscow. It's designed to reflect microwave signals.

The wireless transmitter presents another problem. It has to be detected, and an ordinary RF field strength meter is not enough to do the job. Trouble with a field strength meter is it will measure radio stations in the area as well as any bugs that happen to be nearby.

Ideally, a bugfinder transmits a note or tone which is picked up by the wireless microphone and transmitted. The bug detector then receives the tone that has been transmitted and compares it with the original tone radiated by the bug finder. If it is the same, it locks in on that frequency and helps you to "sniff" out the bug. There are many variations of this anti-bug device but they all rely on basically the same thing—a modulated tone that is picked up by a microphone and transmitted.

If you find a bug, there are several things you can do. The first impulse naturally is to smash the darn thing and put it out of business. But wait, if you smash it, will you be doing yourself any good? A professional will not place one bug, but will place two or three in a given room. This means that where there is one, there should be a couple more, and you may never find them. In fact, the pro purposely places one bug in a fairly obvious place so that you'll smash it and proceed to talk freely in a room that still has a couple of bugs operating.

If you find a bug, assume there are others present and play it for all it's worth. Give out false information and let your eavesdropper's client worry about sorting out fact from fiction. It seldom pays to call in the police on a bug discovery case, since they have no way of knowing who the bug belongs to, and it's impossible to trace through the manufacturers, especially since most such manufacturers are small-time operators who have no identifying symbols or characteristics on the bug. And even if they did, they would probably have no record of the person who purchased the gadget.

Index

A

Acetate, 12
Acoustics, 79, 91
Alkaline batteries, 45
Ampere-hour capacity, battery, 44
Amplifying mixers, 115
"Anti bug," 221
Automatic reversing, 36
Automatic shutoff, 34
Automatic sync, sound track, 171
Automobile recorder, 32

B

Balanced output, microphone, 73
Bandwidth, video recording, 207
Batteries, 43
Battery charger, 55
Battery-operated recorders, 27
"B" batteries, 44
Belts, drive, 182
Boom microphone stand, 84
Braking, 182
Buffer amplifier, 131
"Bug," 216
"Bug finder," 222

C

Cable, coaxial, 137
Cardiod microphone, 69
Cartridge changer, 18
Cartridges, 16
Cassette, 17, 23
Cell, battery, 44
Center-fill microphones, 90, 189
Ceramic microphones, 56
Charger, battery, 55
Cleaning heads, 179
"Close" miking, 76
Closed-circuit jacks, 145
Coaxial cable, 137
Componet system, 155
Condenser microphone, 64
Connectore microphone, 72
Continous-loop cartridge, 18
Counter, index, 33
Crosstalk, 205
Crystal microphone, 56

D

Darlington amplifier, 106
Deck, 29
Degaussing heads, 180
Demagnetizing heads, 180
Desk stand, 83
Diagonal wrap, 209
Diaphragm, microphone, 56
Din plug, 147
Dolby system, 202
Domains, 7
Drive, 38
Dubbing, 31, 129, 165, 199
Duplicating, 129, 165, 199
Dynamic microphone, 57

E

Editing, 191
Editing sound tracks, 170
Electronic eavesdropping, 216
Electronic editing, 192
Electron-ray recording indicator, 41
European plugs, 147

F

Female jacks, 149
Field, magnetic, 11
Film sound stripe, 167
Five-pin plugs, 147
Floor stand, 83
Flux density, 9
FM wireless microphone, 95
Four-input mixer, 115, 122
Frequency response, 10
Frequency response, microphone, 67, 75
Frequency response, wireless microphone, 99

G

Gap, recording head, 10
Gibson Girl splice, 177

H

Head cleaning, 179
Head degaussing, 180
Helical scan, 208
High impedance microphones, 71

I

Impedance, phone recording, 158
Integrated circuit power amplifier, 112
Integrated circuit preamp, 108
Interlace, 211
Iron oxide coating, 12

J

Jack, phono, 141
Japanese plugs, 145
Jumper cable, 147

L

Lavalier microphone, 77
Leader tape, 14, 175
Leclanche cell, 45
Level indicators, 40
Line preamp, 161
Long-distance bug, 219
Low-impedance microphones, 71
Low-print tape, 14

M

Magnetic field, 11
Magnetism, 7
Maintenance, 177
Manganese-alkaline batteries, 45
Master tape, 199, 206
Matched pair, microphone, 73
Matching transformer, microphone, 71
Mechanical reverberation, 185
Mercury batteries, 47
Microphone impedance, 69
Microphone mixer, 85
Microphone placement, 75
Microphone phasing, 79
Microphone types, 56
Microphone stand, 82
Mixer, microphone, 85
Mixers, 115
Motors, 38
Multiple input switching, 150
Multiple-recorder switching, 129

N

Newell transport, 195
Nickel-cadmium batteries, 44, 50
Noise Keller, 202

O

Omega wrap, 210
Omnidirectional microphone, 67, 77
Open-circuit jacks, 146

P

Panpot control, 91, 189
Pause control, 40
Phasing, microphones, 73, 79
Phone plug, 142
Phono cable, 137
Phono jack, 141
Phono plug, 139
Pickup pattern, microphone, 67
Pressure pads, 182
Piezo crystal microphone, 56
"Pop" filter, 79
Portable recorders, 32
Power amplifier, 109
Preamps, 103

R

RCA phono plug, 139
Rechargeable batteries, 47
Remote control, 40, 123
Response, microphone, 67, 75
Response recorder, 10
Response, wireless microphone, 99
Reverberation, 185

S

Scrambler, 221
Shielded phono cable, 137
Shock mount, 83
Shorting-type jacks, 145

T

Tape-to-tape transfer, 129
Telephone amplifier, 161
Telephone pickups, 158
Telephone tap, 216, 218
Tension, braking, 182

V

Velocity microphone, 60, 79
Vibration mount, 83
Video tape recorder, 207

W

Wind screen, 79
Wireless conference broadcaster, 166
Wireless microphone, 95, 216
Wow, 182

Y

"Y" connectors, 153

Z

Zinc-carbon batteries, 44